Hard Math *for* Middle School: Workbook

Glenn Ellison

ISBN-10: 1-5428-3508-9
ISBN-13: 978-1-5428-3508-4

Introduction

Welcome to *Hard Math!* At Hard Math we think that it is fun to see that you can do way more with middle school math than they ever show you in school. We think that it is fun to feel good when you get problems right instead of worrying about possibly getting some problem wrong. You are holding a workbook full of hard problems designed to let you do this.

The problems in this book accompany my textbook *Hard Math for Middle School: IMLEM Plus Edition*. If you don't have a copy, you really should get one. It's way cheaper than a normal textbook. And this workbook isn't very useful without it. The textbook is divided into 104 sections. This workbook has a worksheet of problems covering the material in each of those sections. There are actually 109 worksheets because I thought some sections were too important to get just one worksheet. But the basic idea is that you can read a section of the textbook and then do the corresponding worksheet to see if you understand the material and practice what you learned. Some kids like to do it backwards: do the worksheet first and then read the book to see what I say about how to do them. But read the book either way. If you get the problems right you may indeed know everything I say in the book. But sometimes you'll learn that there was an easier way.

The prose in the *Hard Math* textbook is all about learning how to do well on IMLEM math contests. But the math on IMLEM contests is a lot like the math on any other middle school math contest. You can also think of the book and textbook as a one-year course that teaches everything you need to know to do well on Mathcounts®. I've thought about making up a new *Unauthorized Mathcounts Edition* of the book replacing the IMLEM-focused prose with Mathcounts-focused prose. But I have a job and hardly make any money on these books anyway, so I've never gotten around to it. The book is also good for preparing for the AMC 10. It covers most of what you need to know to do well on the AMC 10, but not everything.

The most important thing to know about this workbook is that the *Hard* in the title is not a joke. The problems this book are all much harder than the problems in regular middle school books. This can be daunting to kids who are used to always knowing how to do every problem their teachers give them. But most find it pretty easy to get into the spirit of it.

To make it easier to find appropriate problems I arrange the problems on each worksheet roughly in order of increasing difficulty and annotate each problem with a difficulty level. (L1) problems would be hard if you haven't read the book, but become just medium to hard once you know the material. (L2) problems are hard even if you know the material. (L3) are really hard. And (L4) problems are really, really hard.

My recommendation to most 6th graders doing IMLEM is that they should focus on just the (L1) and (L2) problems. Even the (L2) problems are often very hard. The (L3) problems will still be hard when you're in 7th and 8th grade, but you'll know the material better when you go through it again. By all means do try the (L3) problems right away if you want. The whole idea of this book is to provide opportunities. (L4) problems only start appearing when you get farther into the book. They are sufficiently hard so that even most 8th graders who are serious about math team will mostly want to skip them. But some have very neat solutions and I thought I should include them especially in the Mathcounts sections because some of my readers are getting ready for state and national level competitions and I wanted to make sure there were things to challenge even them.

I give answers to all questions at the back of the book. Chris Avery and Karthik Seetharaman caught a number of mistakes in the first printing for me, but I'm sure there are more. Chris and I are still working on a solution manual. Including all the solutions would make the printed workbooks too long, so I'm not sure what we'll do with it if it does get finished. By the time you are reading this, hopefully at least part of the solution manual will be available either on my website or on Amazon.

Again, welcome and I hope you enjoy *Hard Math*! If you get through even part of it, you should feel good that you have taken on a real challenge and learned much more than you could have if you had stuck with regular math.

Meet #1 Geometry: 2.1 Basic Definitions

1. (L1) Angle ABC is an obtuse angle. Which of the following could be its measure in degrees: 30º, 60º, 90º, or 120º?

2. (L1) The figure below is drawn to scale. The measures of the three angles of triangle ABC (not necessarily in order around the triangle) are 20º, 40º, and 120º. What is the positive difference between the measure of angle A and the measure of angle C?

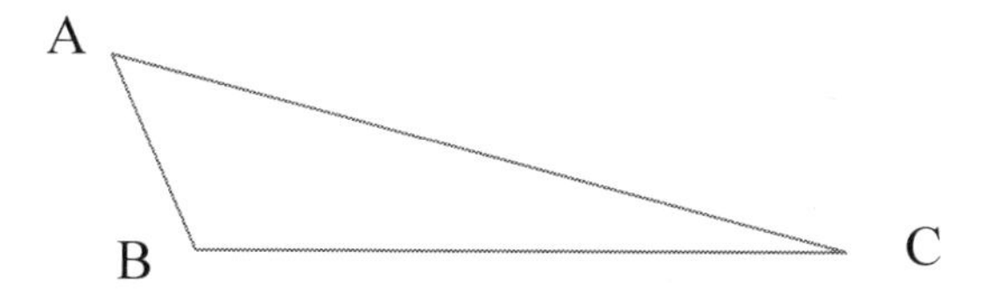

3. (L2) In the figure below the measure of angle ABC is 40º and the degree measure of angle ABD is a multiple of 7. What is the largest possible value for the measure of angle ABD?

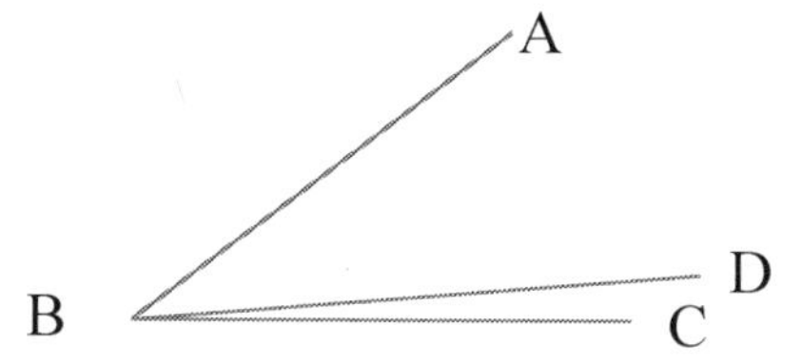

4. (L3) Let ABCD be a quadrilateral. Exactly two of its four interior angles are obtuse angles. The sum of the degree measures of angles DAB and CDA is 200º. The sum of the degree measures of angles BCD and CDA is 210º. One angle in the quadrilateral has a measure of 100º. Angle ABC is acute. What is the degree measure of angle DAB?

Meet #1 Geometry: 2.2 Adding Up Rules

1. (L1) What is the measure of an angle if the measure of its complement is 65º?

2. (L1) In the figure below the measure of angle BOD′ is 155° and the measure of angle BOG is 140°. What is the positive difference between the measure of angle HOG and the measure of angle D′OH?

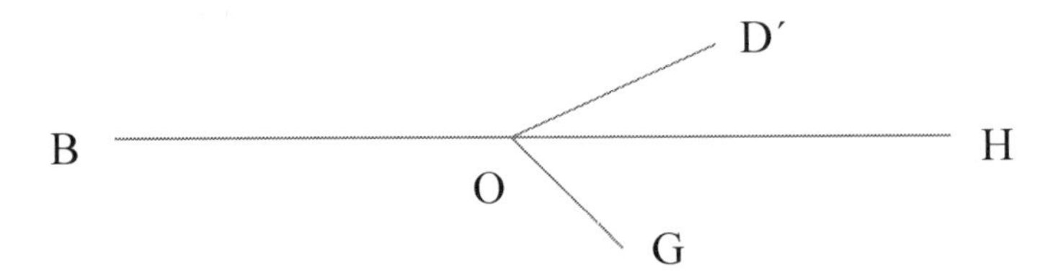

3. (L2) The measure of the supplement of an angle is 20° more than twice the measure of the complement of the angle. What is the measure of the angle?

4. (L2) In the figure below point M is on segment GB and OMC and BMD are right angles. If the measure of angle CMD is 53°, what is the degree measure of angle OMG?

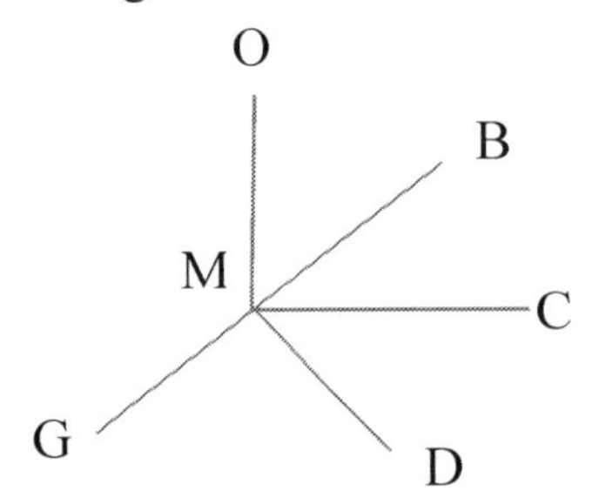

5. (L3) The measure of the supplement of the complement of the supplement of an angle is 7.2° less than the angle's measure. What is the measure of the angle in degrees?

Meet #1 Geometry: 2.3 Equality Rules

1. (L1) Lines m and n are parallel. If the measure of angle DOG is 140°, what is the measure of angle PIG?

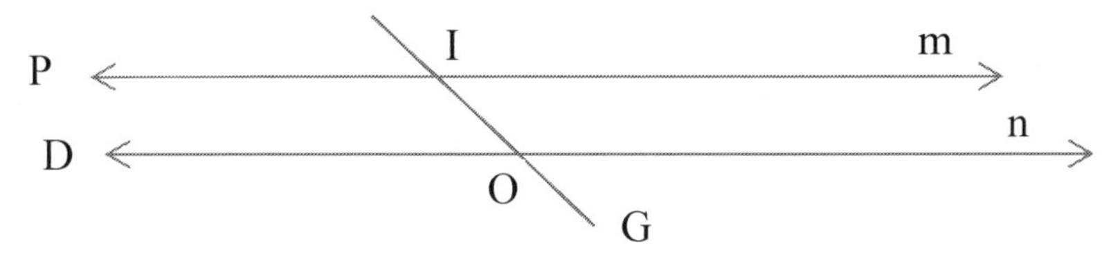

2. (L2) Lines m and n are parallel. The measure of angle COP is 70° and the measure of angle DGF is 48°. What is the degree measure of angle FOA?

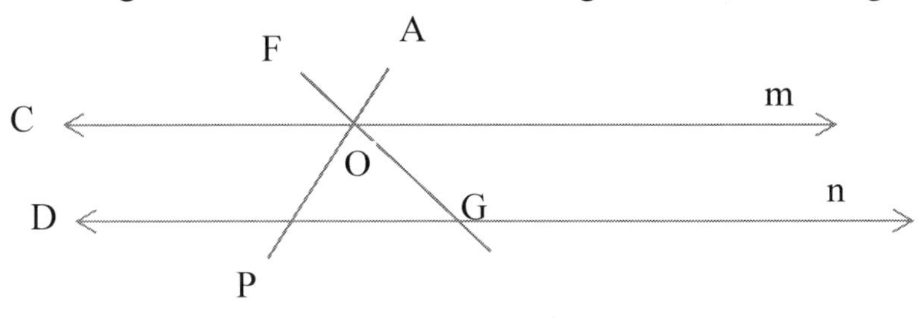

3. (L3) In the figure below lines DX and BG intersect at M and angles OMC and BMD are right angles. The measure of angle CMD is 48°. The measure of angle ABM is 42°. What is the degree measure of angle BEX?

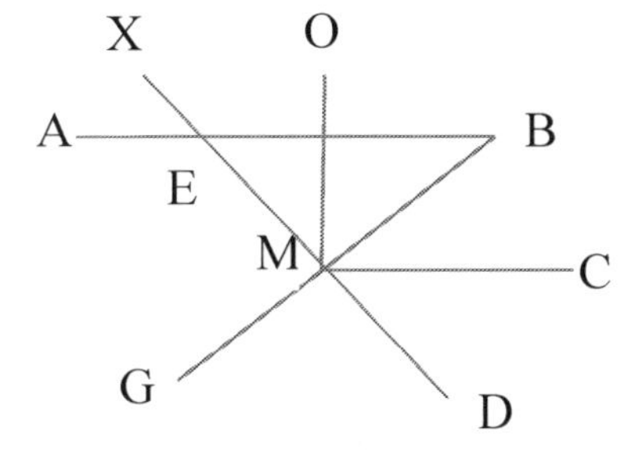

Meet #1 Geometry: 2.4 Angles in Polygons

1. (L1) What is the measure of angle DAB in the figure below?

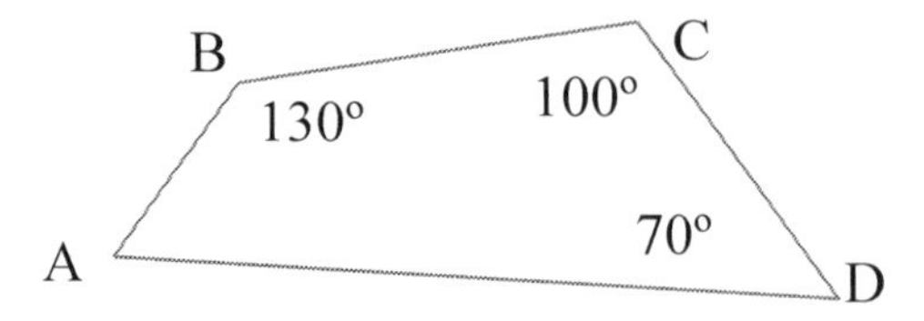

2. (L2) What is the measure in degrees of an interior angle in a regular decagon? (A decagon has 10 sides.)

3. (L3) The exterior angles of a regular hexagon and a regular N-gon are complements. What is N?

4. (L3) In the figure below ABCDE is a regular pentagon, KATNIS is a regular hexagon, and CB is parallel to KS. What is the degree measure of angle EAT?

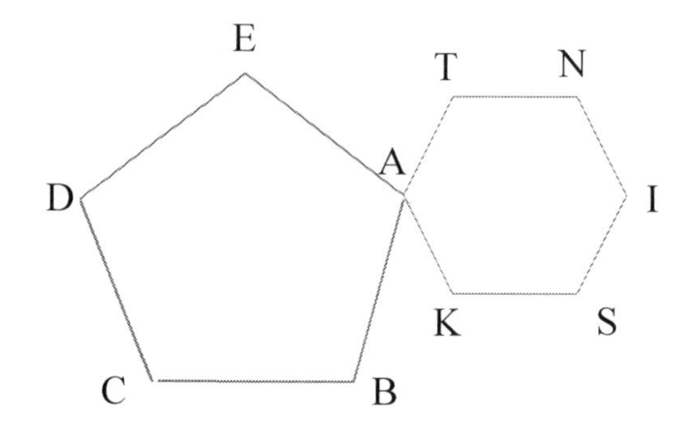

Meet #1 Geometry: 2.5 Problem Solving Strategies

1. (L1) If ABCDE is a regular pentagon and BCNK is a square, what is the measure of angle ABK?

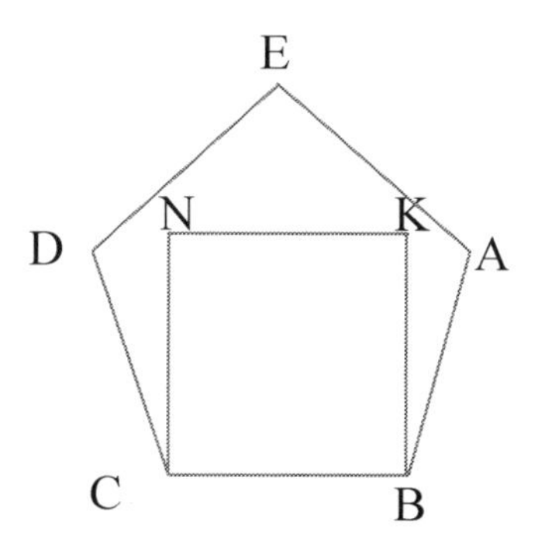

2. (L2) In the figure below points A, B, and C are collinear and DBE is a right angle. If the measure of angle ABD is 141.5927° what is the measure of angle CBE?

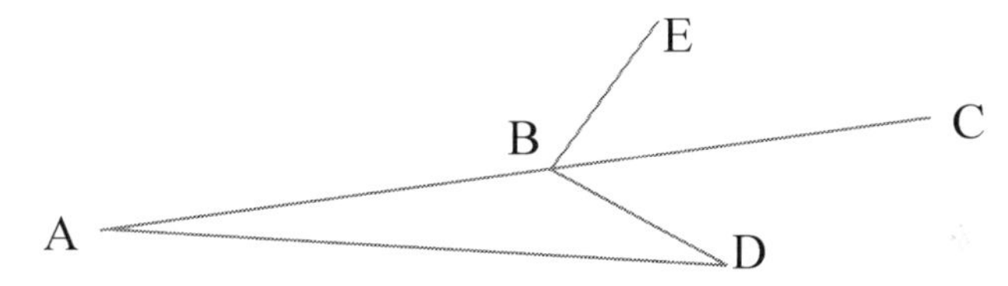

3. (L3) In the figure below, line segment AF intersects CE at D and CB at X. If the measures of BAD and EDF are both 20°, the measure of ABC is 115°, and DFE and DXC are complements, what is the measure of angle AGF?

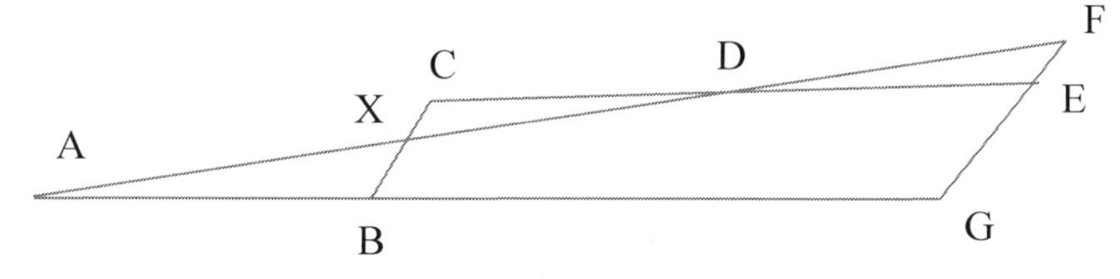

Meet #1 Geometry: 2.6 Advanced Topics

1. (L1) If ABCDE is a regular pentagon, what is the degree measure of angle ECA?

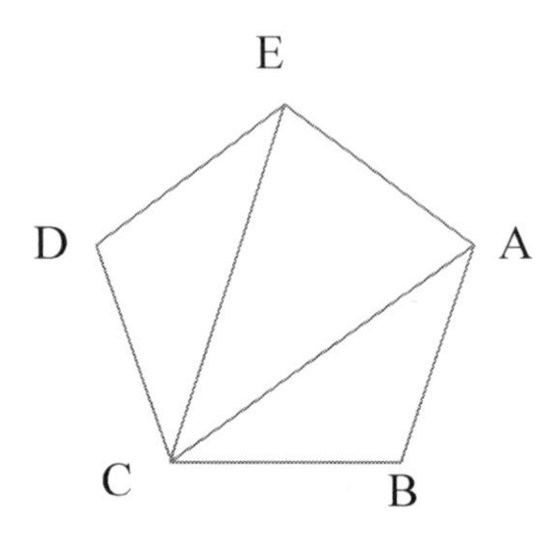

2. (L2) In the figure below ABC and BCD are right angles, CDE is an equilateral triangle, and segments AB, BC, and CD have the same length. What is the measure of angle AED?

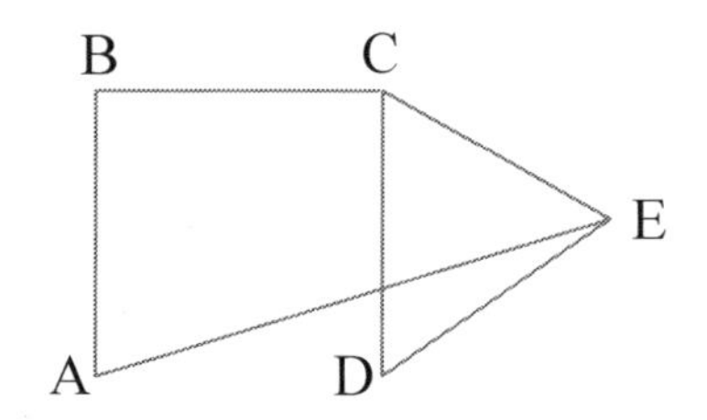

3. (L3) Let ABCD be a rectangle. Let X be the midpoint of AC and let E the point where the internal bisector of angle ABC intersects AC. If AB=XC, what is the median of the degree measures of the angles of quadrilateral DFEX?

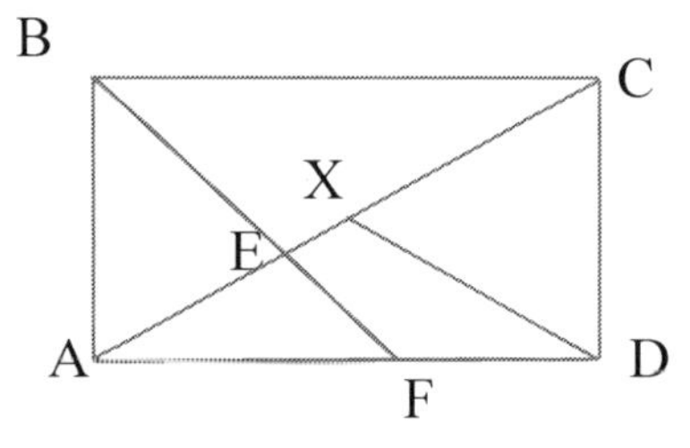

Meet #1 Number Theory: 3.1 Prime Numbers

1. (L1) List all prime numbers between 20 and 30 in order from smallest to largest.

2. (L1) What is the largest prime number that is less than 5×8?

3. (L2) Two prime numbers are called twin primes if they differ by two. What is the largest pair of twin primes that is less than 40?

4. (L2) What is the average of all composite numbers that are between 30 and 50? (Do not count 30 or 50).

5. (L2) The numbers 11, 13, 17, and 19 are all prime. Hence, the answer to the question of what is the smallest number x for which $10x + 1$, $10x+3$, $10x+7$ and $10x+9$ are prime is 1. What is the second smallest positive integer x with this property?

6. (L3) While Anna was reading a book she announced, "Hey, the number of the page I'm on is a product of two primes which differ by 6." Caroline replied, "I've read more pages than you and the same is true of the page I'm on." Kate replied "Be quiet, I'm trying to listen to "We Are Golden."

What is the smallest possible number of the page that Caroline could be reading?

Meet #1 Number Theory: 3.2 Prime Factorization

1. (L1) Find the sum of the prime factors of 22.

2. (L1) What is the smallest prime factor of 204?

3. (L2) Let a, b, c, and d be whole numbers with $108 = 2^a \times 3^b \times 5^c \times 7^d$. Find $a + b + c + d$.

4. (L2) Suppose that $2700 = 2^a \times 3^b \times 5^c \times 7^d \times 11^e$ and a, b, c, d, and e are all whole numbers. Find $a \times b + c \times d$.

5. (L2) Let a, b, c, d, e, and f be whole numbers with $555{,}000 = 2^a \times 3^b \times 5^c \times 7^d \times e^f$. Find $a + b + c + d + e + f$.

6. (L3) What is the largest prime factor of the sum of the prime factors of the product of the prime factors of the product of the prime factors of 105?

7. (L3) Let s be the number of seconds in the year 2015. Let m be the number of minutes in a day, and let $y = s \div m$. Write $y = 2^a \times 3^b \times 5^c \times 7^d \times e^f$ where a, b, c, d, e, and f are all whole numbers. Find $a + b + c + d + e + f$?

8. (L4) Suppose that the positive integer x can be written as $2^a \times 3^b \times 5^c$ and $x+1$ can be written as $2^d \times 3^e \times 5^f$. Suppose that a, b, c, d, e, and f are all whole numbers and that $a + b + c > 3$. What is the smallest possible value of x?

Meet #1 Number Theory: 3.3 Counting and Summing Factors

1. (L1) List all factors of 75 in order from smallest to largest.

2. (L1) How many factors does 36 have?

3. (L2) How many factors does 1440 have?

4. (L2) What is the sum of the factors of 200?

5. (L2) What is the third largest factor of 225?

6. (L3) How many factors of 1935 are perfect squares?

7. (L3) What is sum of all of the odd factors of 1080?

8. (L4) Find the sum of the squares of all factors of 726.

Meet #1 Number Theory: 3.4 Divisibility Rules

1. (L1) For what values of A is the five-digit number 31A2A a multiple of five?

2. (L1) For what value of B is the five-digit number 284B2 a multiple of nine?

3. (L2) For what value of A is the five digit number 31A2A a multiple of 12?

4. (L3) The five-digit number 35A2A is divisible by 2, 3, 4, 6, 8, 9, and 12. What is the value of A?

5. (L3) If AB9BA is divisible by 72, what is the value of A + 2B?

6. (L4) Find all values of A for which 3A524 is a multiple of 37.

7. (L4) Find all ordered pairs (A,B) for which 6A335B is a multiple of 74.

8. (L4) Find all values of A for which the base 63 number $A7894321_{(63)}$ would end with a zero if it were written in base 10.

Meet #1 Number Theory: 3.5 Problem Solving Tips

1. (L1) How many two digit odd numbers are perfect squares?

2. (L2) Let n be a two-digit positive integer. Find the value of n if
   - The tens digit of n is smaller than the ones digit of n.
   - n is a perfect square.
   - n is two more than a prime number.
   - n is a multiple of 5.
   - The sum of the digits of n is a prime number.

3. (L2) Find the value of n if
   - n is a positive integer
   - $10 \times n + 19$ is prime.
   - n is a factor of 15.
   - n is a factor of 56,595,000.

4. (L3) How many numbers less than 100 do not have a 2, 3, or 5 in their prime factorization?

5. (L3) What is the smallest number that is a perfect square, a perfect cube, and has 49 factors?

6. (L3) Find the value of n if
   - n is a positive integer
   - $10^n - 1$ is a multiple of 7.
   - n is not a multiple of 17.
   - n is a three digit number.
   - The hundreds digit of n is greater than the tens digit.
   - n is a perfect cube.

Meet #1 Arithmetic: 4.1 Order of Operations

1. (L1) Evaluate the following expression as a decimal to the nearest thousandth.

$$\frac{3+4\times 5+2^{1+1}+1}{2^{3+4\div 2}}$$

2. (L1) Evaluate the following expression.

$$\frac{3^{2+1}+(-2\times(5+3))-3\div 3}{44\div 22\times 3-5\div 5}$$

3. (L2) Evaluate the following expression as a decimal to the nearest hundredth.

$$\frac{146\times 6+(5-5\times 145)-2^{0}}{17^{2}-14^{2}}$$

4. (L3) Evaluate the following expression.

$$\frac{2^{\frac{4+3}{2}}-4^{\frac{2\times 3}{4\cdot(4+4)\div 4}}}{2^{2^{-1}}}$$

Meet #1 Arithmetic: 4.2 Statistics

1. (L1) Find the average of 397, 810, 596, 1011, and 188.

2. (L1) Ms. Mercado's class has 12 girls and 9 boys. On the first test the girls averaged 91 and the boys averaged 84. What was the class average?

3. (L2) The stem and leaf plot below shows the scores earned by Ms. Gao's class on a test. What is the positive difference between the mean score and the median score?

Stem	Leaf
9	0 0 2 5 9
8	2 5 7 8
7	4 6 6 8 8

4. (L2) Krishna's scores on his math tests last term were 79, 87, 90, 92, and 99. One of the tests was more important than the others and was counted twice in computing his average for the term. His teacher gave anyone with an average of at least 90 and less than 93 an A- and gave anyone with an average of at least 93 an A. What was Krishna's score on the test that counted twice if he got an A- for the class?

5. (L3) Julia had five quizzes in her biology class. They were scored on a 0 to 100 scale and all scores were whole numbers. Her scores on three of the tests were 84, 88, and 95. What is the difference between the largest possible value for her median score and the smallest possible value for her median score if her average was 90?

Meet #1 Algebra: 5.1 Simplifying Expressions

1. (L1) Simplify $3(2x - 2) + (x + 5)$.

2. (L1) Simplify $5(2 - x) + 5x + 16(2 - x)$.

3. (L2) Simplify $3x - 2 + 187(2x - 5) + 3x + 2 + 13(2x - 5)$

4. (L2) Simplify $313(2x - 7) + 13(7x - 5) + 313(5x + 7) - 7(13x - 5)$

5. (L2) Simplify $3(3 - 2x) + 27(137x - 3) - 3(2x - 3) - 27(136x - 4) + 5$

6. (L3) Simplify

$81(25x + 1) - 5(x - 99) - 45(45x + 11) + 9(x - 9) + (2025x - 243) \div 9^2$.

Meet #1 Algebra: 5.2 Evaluating Expressions

1. (L1) Evaluate $3(7x - 2) - 20x$ for $x = 17$.

2. (L1) Evaluate $10 - 5x + 4x - 3 + (2 - 5x)$ for $x = 1/6$.

3. (L2) Find $(3x + 17) + 21(x - y) + 5(4y - 3) - 14x$ if x is ¼ and y is 3½?

4. (L2) Evaluate the following expression for $x = 1/7$.

$$37(2x - 5) + 7(7x - 5) + 35(5 - 2x) - 6(8x - 6) + (x + 9)$$

5. (L3) What is the value of the expression below if $x=120/17$ and $y=60/17$? Write your answer as a mixed number.

$$131(x - 5y) + 508(x - y) + 129(x + y) + 500(x - 3y)$$

6. (L3) Ankur evaluated the expression $137(2x - y) + 79(3x + 4y)$ for $x=22/7$ and $y = 1/2$. Mengxi evaluated the expression $135(2y - x) + 80(4x + 3y)$ for $x=1/2$ and $y=22/7$. If they both did their problems correctly, how much larger was Ankur's answer?

Meet #1 Algebra: 5.3 Solving Equations in One Unknown

1. (L1) Solve for x: $5x - 13 = 12$.

2. (L1) For what value of x does $12 - 3x = 2x + 3$? Write your answer as a decimal.

3. (L2) Find x if $3(2 + 5x) - (x + 4) = 3x - 1 + 2(5x + 2)$.

4. (L2) Carter likes to eat well-balanced junk-food snacks, so on his most recent trip to the 7-11 he bought 3 Twix bars, a 99 cent fountain Coke, and a 59 cent package of Cheetos. If he paid with a $10 bill and got $6.05 in change, how much does a Twix bar cost?

5. (L3) Find the value of x that solves the equation below. Express your answer as a common fraction in simplest terms.

$$87\,(5x + 17) + 23(16 - (5x + 17)) = 8\,(2 + 6\,(5x + 17))$$

6. (L3) Dong Gil is three years younger than Dong Yeop. In 7 years Dong Jae will be three years older than Dong Yeop was last year. If the mean of Dong Gil, Dong Yeop and Dong Jae's ages is $14\frac{1}{3}$, how old is Dong Gil?

Meet #1 Algebra: 5.4 Identities

1. (L1) For what value of W is the expression $3(x-2) = 3x - W$ an identity?

2. (L1) For what value of A is the expression $A(x+3) - 2(x-5) = x + 19$ an identity?

3. (L2) Find K if $18(2x+37) + 6((K-2)x + 55) = 25(x+39) + 21$ is an identity.

4. (L2) For what value of W is the equation below true for any number x?

$$3(3x + 2W - 5) + 4(8x + 2W)/2 = 5(5x + W)$$

5. (L3) For what value of A is the equation below an identity?

$$(3A-2)(237x-1) + (4-5A)(121x-2) = (6+A)(105x+1) + 6(A-2)$$

6. (L3) For what value of B is the equation below true whenever $x = y$?

$$2((3+B)x - y + 1) + 3Bx + 5(x - (4-B)y + 3) = -y + 17$$

Meet #1 Algebra: 5.5 Made-up Operations

1. (L1) If A⌂B is defined to be (A + B) × A, what is ½⌂½?

2. (L1) If x☼y means $x^2 - y^2$, what is (5☼4)☼8?

3. (L2) If x☺y means $(x^2 - y) \times y$ and x ☻ y means the positive square root of (x – y) what is (5☺2) ☻ 10 – (21 ☻ 5)?

4. (L2) If x♥y means $(x^2 + x - 2y)$ what is 0♥(0♥(0♥ (3♥-2)))?

5. (L3) If x♦y means to find the sum of the factors of x × y, what is (11♦11) ♦101?

6. (L3) If x☺y means $x^2(x^2+y)$ what is the median of (3☺2)☺1, (2☺3)☺1, (1☺2)☺3, 3☺(2☺1), and (1☺(2☺3)?

Meet #2 Geometry: 2.1 Perimeter

1. (L1) Find the perimeter of triangle ABC if the distance between the grid lines in the figure below is 1cm?

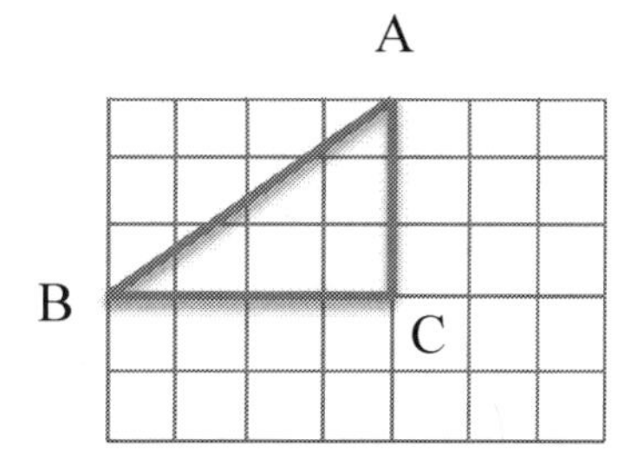

2. (L2) Find the perimeter of hexagon ABCDEF if the distance between the grid lines in the figure below is $\sqrt{5}$ units? Write your answer in the form $a + b\sqrt{5}$ where a and b are whole numbers.

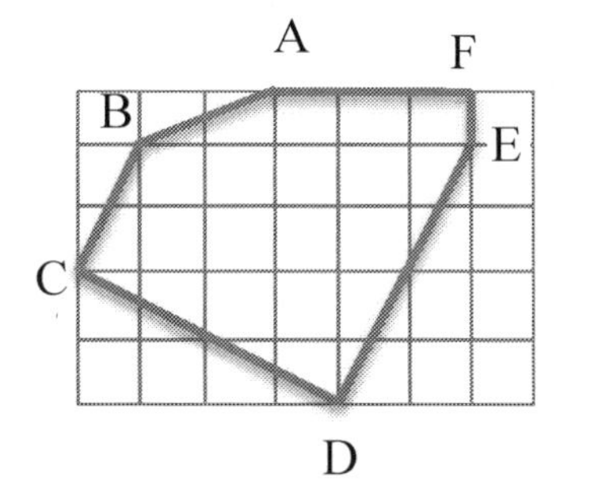

3. (L3) In the figure below ABCDEF and GELMON are regular hexagons with side length 6cm and the measure of angle DEL is 60°. What is the perimeter of pentagon ACEOG in centimeters?

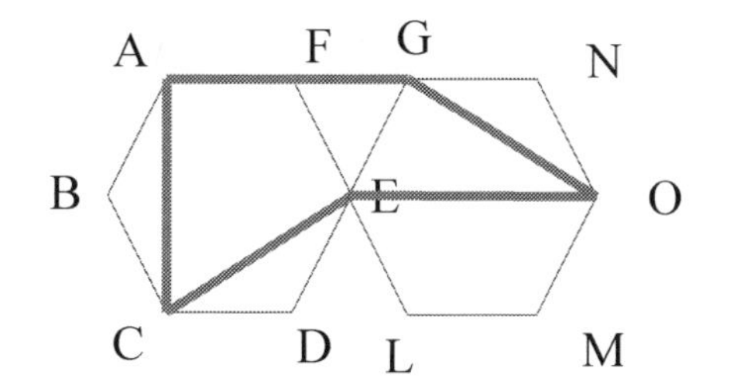

Meet #2 Geometry: 2.2 Areas

1. (L1) Distances between stars are very big if measured in miles, so astronomers sometimes measure distances in parsecs. (One parsec is a little more than 19 trillion miles. The second closest star to the earth is about 1.3 parsecs away.) What is the area in square parsecs of the triangle below?

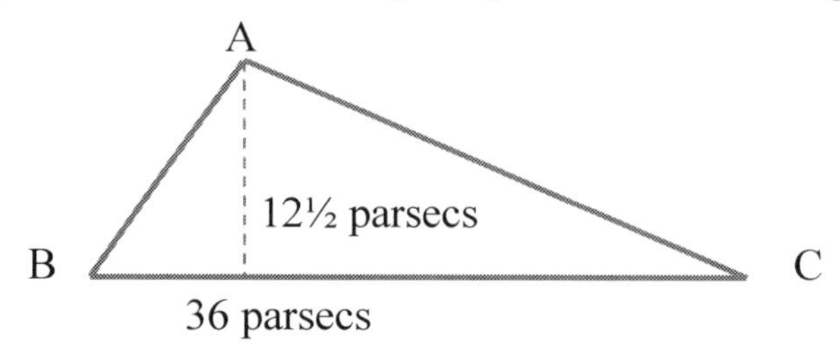

2. (L2) Helen figured out that she could draw a regular octagon with side length 2 cm using a three step procedure. She first draws a square with side length 2 cm. She then adds four rectangles which are 2 cm in one direction and $\sqrt{2}$ cm in the other. Finally she connects the corners of her figure. What is the area in square cm of her octagon?

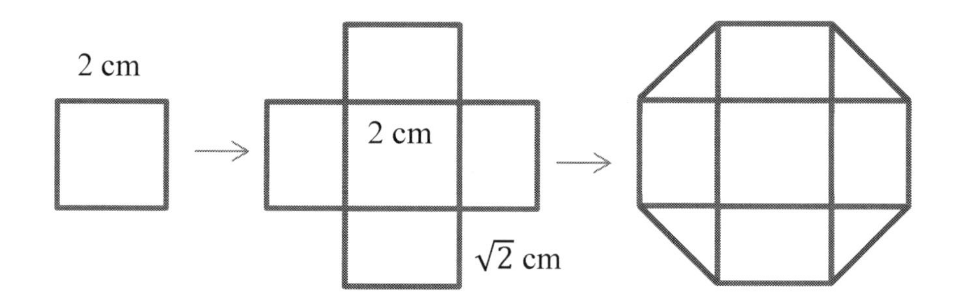

3. (L3) What is the area in square feet of the trapezoid shown below?

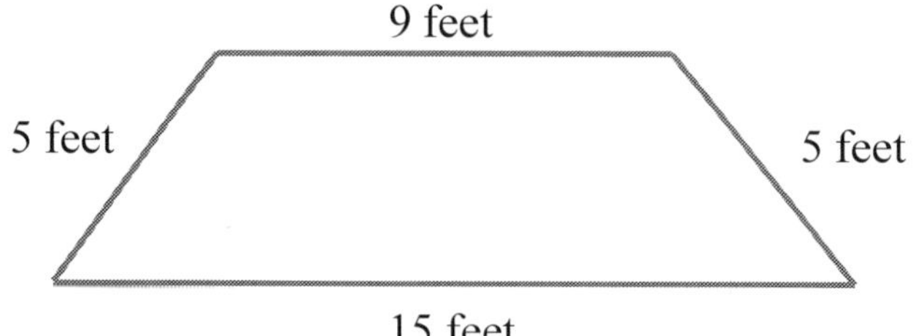

Meet #2 Geometry: 2.3 Perimeters of Rectilinear Figures

1. (L1) The map below shows the city in which Wei lives. Each block in her city a square that is 500 feet on a side. She walks from A to B following the path marked by a solid line. She then walks back to A via the point marked C following the dashed lines. How far does she walk in total?

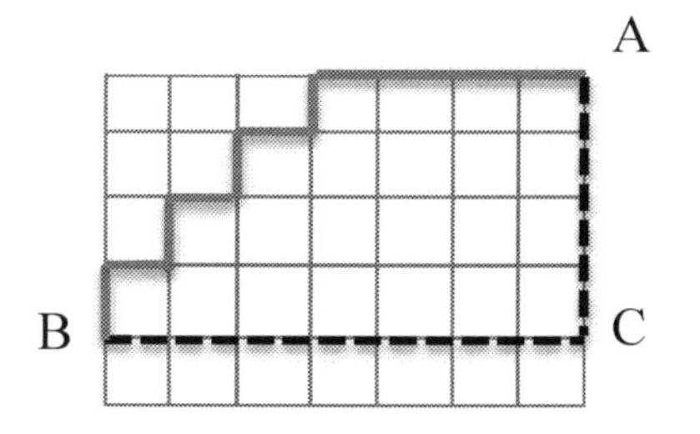

2. (L2) Find the perimeter of figure below if all sides meet at right angles.

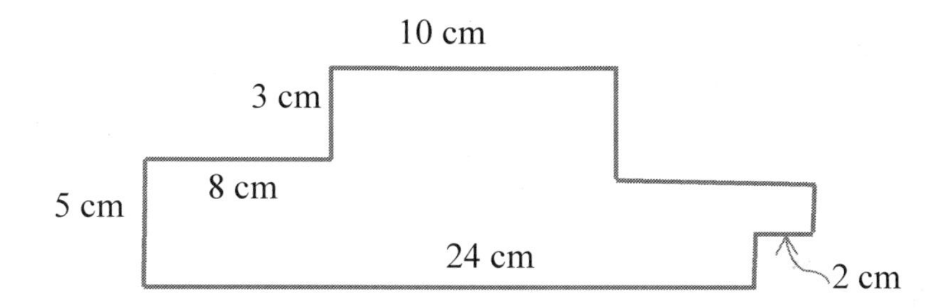

3. (L3) Find the perimeter of the hexadecagon shown below if its area is 570 square cm and all of the angles are right angles.

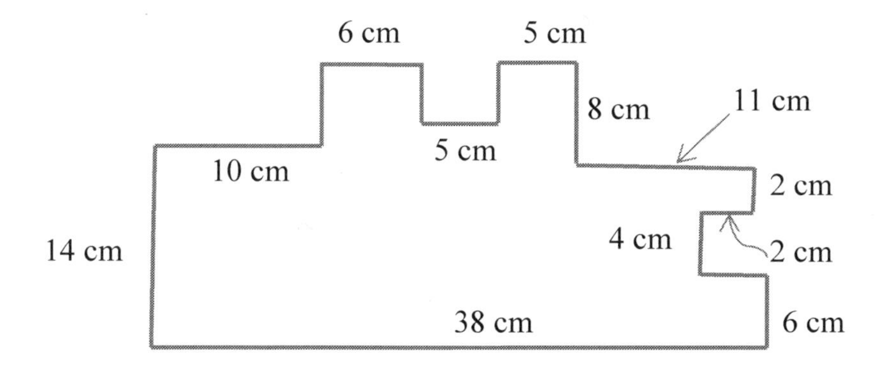

Meet #2 Geometry: 2.4 Problem Solving Tip: Subtract

1. (L1) Normally we think of the diagonals of a quadrilateral as intersecting inside the quadrilateral. But in a nonconvex quadrilateral like ABCD below, the point H where diagonals BD and AC intersect is outside the quadrilateral. Find the area of ABCD in cm^2 if the length of BD is 4 cm, the length of DH is 5 cm, the length of AH is 5 cm, the length of HC is 11 cm, and BD and AC are perpendicular.

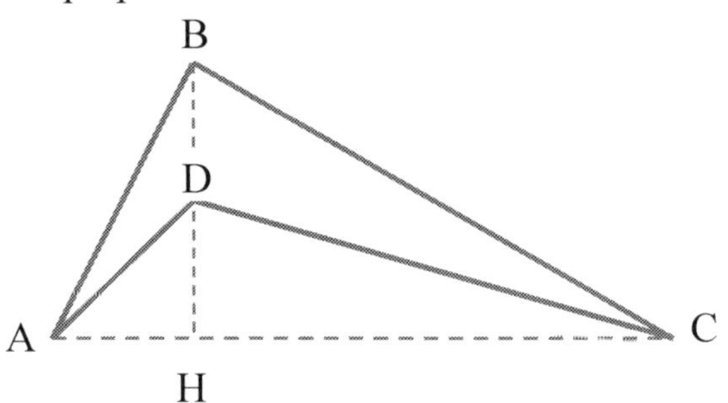

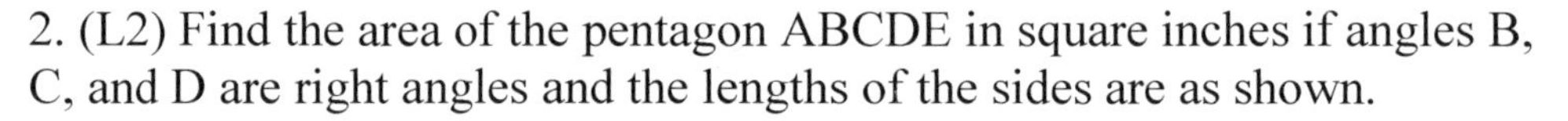

2. (L2) Find the area of the pentagon ABCDE in square inches if angles B, C, and D are right angles and the lengths of the sides are as shown.

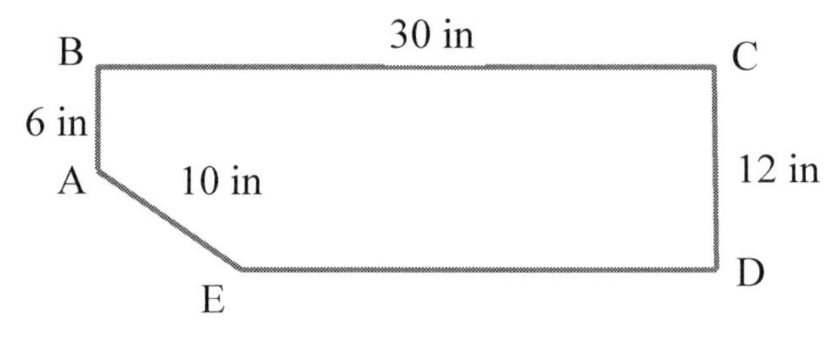

3. (L3) What is the area in square feet of triangle ICE if the grid lines in the rectangular grid shown are 6 inches apart? Give your answer as a simplified mixed number.

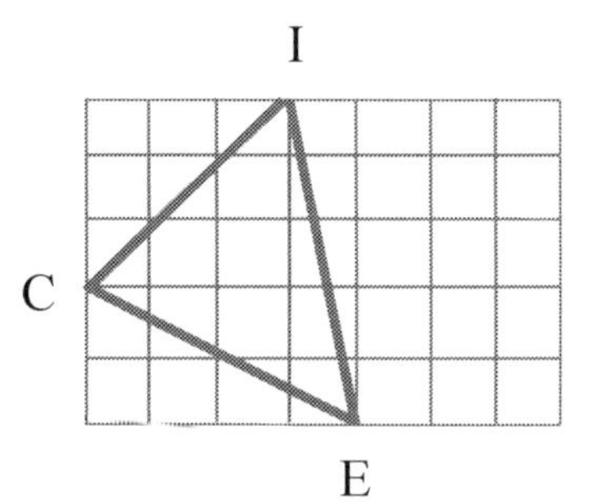

Meet #1 Geometry: 2.5 Advanced Topic: More Area Formulas

1. (L1) What is the area in cm^2 of the triangle below?

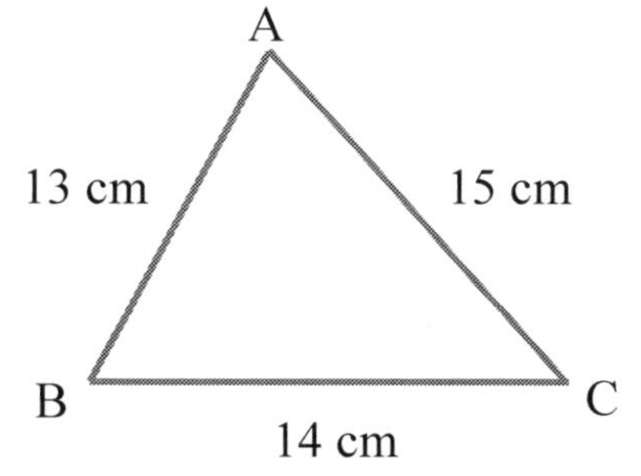

2. (L2) A circle is inscribed in a triangle with side lengths 4 cm, 5 cm, and 7 cm. What is the radius of the circle?

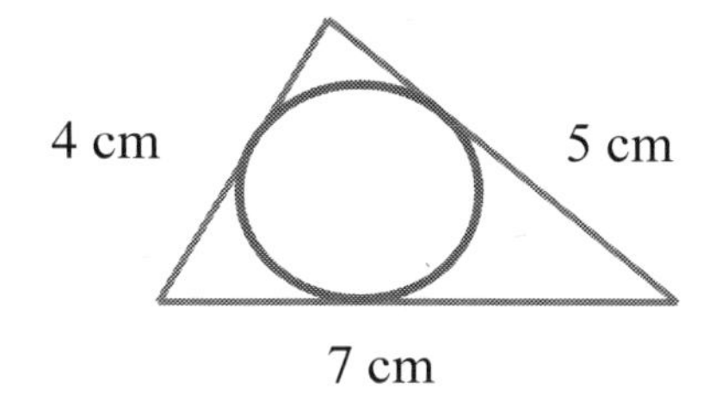

3. (L3) What is the area in square feet of a regular octagon that is inscribed in a circle of radius one foot? Give your answer as a decimal to the nearest hundredth.

4. (L4) What is the are in square feet of the trapezoid shown below?

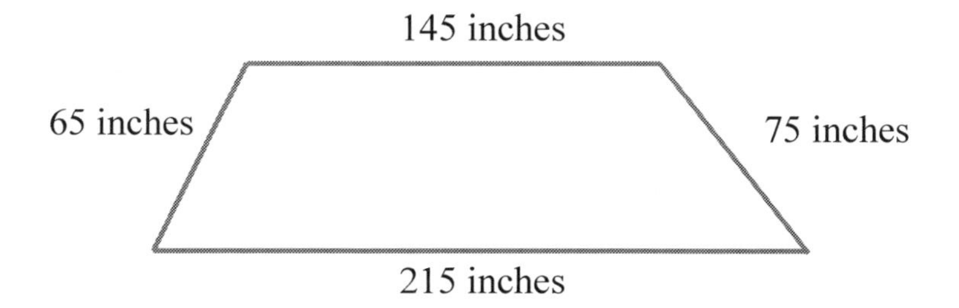

Meet #2 Number Theory: 3.1 A Super-Quick Review of Prime Factorization

1. (L1) Find the sum of the prime factors of 55.

2. (L1) List all factors of 169 in order from smallest to largest.

3. (L2) Let a, b, c, and d be whole numbers with $3500 = 2^a \times 3^b \times 5^c \times 7^d$. Find $(a + b) \times (c + d)$.

4. (L2) Let a, b, c, and d be whole numbers with $980{,}000 = 2^a \times 3^b \times 5^c \times 7^d$. Find the maximum of a, b, c, and d.

5. (L3) Write the square root of 1,464,100 in the form $p^a \times q^b \times r^c$ where p, q, and r are primes arranged in increasing order and a, b, and c, are whole numbers.

6. (L3) Angie was born in the 2000's. She wrote down the date, month (as a number from 1 to 12) and year (as a four-digit number) of her birthday and noticed that none was a multiple or factor of 3, 5, or 7. What is the earliest date on which she could have been born?

Meet #2 Number Theory: 3.2 Greatest Common Factors

1. (L1) What is the largest number that is a factor of both 27 and 39?

2. (L1) Find the greatest common factor of 84 and 119.

3. (L2) What is the greatest common factor of 34,300 and $2^1\ 3^2\ 5^3 7^4\ 11^5$?

4. (L2) The math team bakes 54 yellow cupcakes and 125 chocolate cupcakes for a bake sale. They buy boxes that hold multiple cupcakes and fill the boxes with the same number of cupcakes in each. Every box either has all yellow cupcakes or all chocolate cupcakes. They use up all the yellow cupcakes, but two chocolate cupcakes were left over so the math team coach ate them. How many cupcakes were in each box?

5. (L3) Mae always runs laps of her school track at exactly the same speed. She runs a whole number of laps each day and each lap takes a whole number of seconds. On Monday she runs for 1 hour, 9 minutes, and 31 second. On Tuesday she is tired so she stops after 16 minutes and 10 seconds. What is the longest possible time that it could take her to run one lap?

6. (L3) Ms. Qian's classroom is 15 feet wide and 22½ feet long. The floor is currently tiled with 1 foot by 1 foot tiles. As a result, the tiles on the end are cut in half. Suppose that you want to retile the floor using square tiles arranged in a rectangular pattern, that you want the sides of the tiles to be a whole number of inches long, and that you want to tile the room without cutting any tiles. What is the smallest number of tiles that you could use?

7. (L4) Find GCF(9282,GCF(104101, 22303))

Meet #2 Number Theory: 3.3 Least Common Multiples

1. (L1) What is the least common multiple of 30 and 40?

2. (L1) 22 and 10 are both factors of x. What is the smallest possible value of x?

3. (L2) Find LCM(1212, 420).

4. (L2) Sora and Ji start running laps of the school track at exactly 3:00pm. Sora takes 2 minute and 45 seconds to run each lap and Ji takes 3 minutes and 15 seconds. They run until the first time that they reach the starting line at exactly the same time. At what time do they stop?

5. (L3) Callista put the fractions $\frac{13}{778}$ and $\frac{11}{78}$ over a common denominator so that she could add them together. What is the smallest possible denominator she could have used?

6. (L4) How many two-digit numbers x are solutions to

$$\text{LCM}(x, \text{LCM}(36100, 6460)) = \text{LCM}(36100, 221)$$

Meet #2 Number Theory: 3.4 More on GCFs and LCMs

1. (L1) What is the greatest common factor of the set {14, 18, 26, 21, 20}?

2. (L1) $a \times b = 2688$ and GCF(a,b)=8. Find LCM(a,b).

3. (L2) Two integers x and y satisfy that GCF(x,y)=37 and LCM(x,y)=666. What is $x \times y$?

4. (L2) Bella, Max, Bailey, and Lucy are all dogs that don't particularly like chasing tennis balls. Bella chases every 12th ball thrown, Max chases every 15th ball thrown, Bailey chases every 6th ball thrown, and Lucy chases every 20th ball thrown. If they all chased after a ball at 10:27 and one ball is thrown every minute, when will they next all chase the same ball?

5. (L3) For how many n is GCF(n,6)=2 and LCM(n,60)=60?

6. (L3) If LCM(a,b)=488 and LCM(c,d)=201, find LCM(a,b,c,d).

Meet #2 Number Theory: 3.5 A Slightly Longer Review of Prime Factorization

1. (L1) $230685000=2^3\times3^1\times5^4\times7^1\times13^3$. How many factors does 230685000 have?

2. (L1) What is the prime factorization of 396?

3. (L2) How many factors does 84 have?

4. (L2) What is the sum of factors of 120?

5. (L3) What is the sum of all n such that LCM(n, 48)=48?

6. (L3) Let n=8972058000. What is GCF(n, 22500)?

Meet #2 Arithmetic: 4.1 Fractions and Percents

1. (L1) What is 40% of 85?

2. (L1) What number is 25% greater than two-thirds of 78?

3. (L2) Mr. Smith's class correctly solved 87.5% of the problems they attempted. Mrs. Smith's class correctly solved all but 15% of the problems they attempted. If Mrs. Smith's class attempted 80 problems and Mr. Smith's class attempted 10% more problems, how many more problems did Mr. Smith's class correctly solve?

4. (L2) What number is 35% less than three-elevenths of 55% of 408? Give your answer as a decimal.

5. (L3) A Tesla weighs 2475 pounds. If it were twenty-two and two-ninths percent lighter, then it would be 12.5% lighter than a Toyota Prius would be if the Prius were 10% heavier than four-fifths of its actual weight. How many pounds does a Prius weigh?

6. (L4) The Sodor public schools use the smallest number of teachers in a grade that makes the average class size less than 27. Mae does a survey of the nine 7th grade classrooms and finds (rounding to the nearest whole numbers) that 51% of the students are girls, 33% of the girls wear makeup, and 16% of the 7th graders wear makeup. If no boys in the school wear makeup, how many 7th grade girls are there?

Meet #2 Arithmetic: 4.2 Terminating and Repeating Decimals (part 1)

1. (L1) Write 0.454545… as a fraction in lowest terms.

2. (L1) What is $0.\overline{3} - 0.1\overline{6}$. Express your answer as a fraction in lowest terms.

3. (L2) Write $0.1\overline{2}$ as a fraction in lowest terms.

4. (L2) Simplify $0.1\overline{27}$ / $0.\overline{27}$. Write your answer as a fraction in lowest terms.

5. (L3) Write the difference between $0.10\overline{101}$ and $0.\overline{10}$ as a fraction in lowest terms.

6. (L3) Write the difference between $0.10\overline{101}$ and $0.\overline{10}$ as a decimal using repeating decimal notation.

Meet #2 Arithmetic: 4.2 Terminating and Repeating Decimals (part 2)

1. (L1) Write 1/30 as a decimal using repeating decimal notation.

2. (L1) What is the 20th digit to the right of the decimal point when 5/120 is written as a decimal?

3. (L2) What is the 55th digit to the right of the decimal point when 17/101 is written as a decimal?

4. (L2) Write 3/41 as a decimal using repeating decimal notation.

5. (L3) What is the 20th digit to the right of the decimal point when $\frac{1}{35}+\frac{1}{99}$ is written as a decimal?

6. (L3) What is the 15th digit to the right of the decimal point when 3/13 is written as a decimal?

7. (L4) What is the smallest positive whole number n for which the decimal expansion of 1/n contains the digits 0, 1, 2, 3, 4, 5, 6, 7, 8, and 9?

Meet #2 Algebra: 5.1 Sums of Arithmetic Sequences

1. (L1) Find 15 +25 + 35 + 45 + 55 + 65 + 75 + 85 + 95.

2. (L1) Artur is five years younger than one of his brothers and ten years younger than his other brother. If the sum of the three brothers' ages is 48, how old is Artur?

3. (L2) Elaine posted a video on VTube on a Sunday. On each day after the first it was viewed 101 more times than it was viewed on the day before. By the end of the seventh day it had been viewed a total of 3619 times. How many times was it viewed on its first day?

4. (L2) Jenny added the dates of all of the Sundays in a month. If the total was 75, what day of the week was the 17th of the month?

5. (L3) The sum of 13 consecutive positive multiples of 21 ends with a zero. What is the smallest possible value for the smallest of the multiples of 21?

6. (L3) Solve for x:

$$(x - 8092) + (2x - 8192) + (3x - 8292)$$
$$= (19x + 8142) + (4x + 8167) + (14x + 8192) + (9x + 8217) + (24x + 8242)$$

Meet #2 Algebra: 5.2 Reasoning in Number Sentences

1. (L1) The sum of two times a number and five is equal to nine. Find the number.

2. (L1) Two times the sum of a number and five is one less than ten. Find the number.

3. (L2) Thirteen times the difference when a number is subtracted from eight is one more than one-fourth of five times the sum of the number and twenty five and two fifths. What is the number?

4. (L2) If a number is increased by seven-and-a-half and then multiplied by three, the answer is the same as if the number were first multiplied by three and then increased by seven-and-a-half and then multiplied by one-and-two-thirds. What is the number?

5. (L3) Three-sevenths of 37.5% more than the sum of the result of subtracting a number from four-and-four thirds and one is one quarter of the sum of 60% of the number and one seventh. What is the number?

6. (L3) If a number is multiplied by two-thirds and then increased by two and then multiplied by three-fourths and then has two subtracted from the result, and the same series of operations is then applied to the result, and then applied again to the new result, the result is one less than two more than three less than four more than five less than six more than seven. What is the number?

Meet #2 Algebra: 5.3 Working with Formulas

1. (L1) The Pythagorean theorem says that if a right triangle has legs of lengths a and b, then the length of the hypotenuse is $c = \sqrt{a^2 + b^2}$. What is the length of the hypotenuse of a triangle with legs of length 8 and 15?

2. (L1) The volume of a cone is $V = \frac{1}{3}\pi r^2 h$ where r is the radius of its base and h is its height. What is the height of a cone if its base is a circle with radius 6 inches and its volume is 180π cubic inches?

3. (L2) The surface area of a sphere of radius r is $4\pi r^2$. What is the surface area in square centimeters (to the nearest whole number) of a sphere with radius 15 cm? (The value of π is 3.14159265358979323… .)

4. (L2) If a triangle has side lengths a, b, and c, then a formula for its area is $A = \sqrt{s(s-a)(s-b)(s-c)}$, where $s = \frac{a+b+c}{2}$. Find the area of a triangle with side lengths 5, 5, and 6.

5. (L3) A formula for converting Fahrenheit to the Celsius temperatures is $C = \frac{5}{9}(F - 32)$. For how many temperatures are the Fahrenheit and Celsius temperatures both positive two-digit whole numbers?

6. (L3) The area of a triangle is $A = \frac{abc}{4R}$, where a, b, and c are the side lengths and R is the radius of the circumscribed circle. Another area formula is $A = rs$, where r is the radius of the inscribed circle and $s = \frac{a+b+c}{2}$. Euler's formula for the distance d between the centers of the inscribed and circumscribed circles of a triangle is $d^2 = R^2 - 2Rr$. Find the distance between the centers of a triangle's inscribed and circumscribed circles if the side lengths are 5, 5, and 8, and the area is 12. Give your answer as a common fraction in simplest terms.

Meet #2 Algebra: 5.4 Word Problems with One Unknown

1. (L1) On Monday Ella watched YouTube videos instead of doing her homework. On Tuesday she watched one less than twice the number of videos that she watched on Monday. If she watched a total of 50 videos on the two days, how many videos did she watch on Monday?

2. (L1) Alan ate five more cookies than Nala. If together they ate 13 cookies how many cookies did Nala eat?

3. (L2) Shyam is two years older than his sister and four years younger than his twin brothers. If the average of the four siblings' ages is 13.5, how old is Shyam?

4. (L2) Find the value of ♥ if

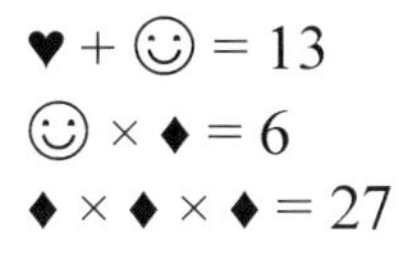

5. (L3) Ms. Rubinstein-Tuchmayer's A block class is her biggest. Her C-block class is the smallest and worst – no student in that class got an A. She gave 7 more A's in her A-block class than in her B-block class. The number of B's she gave in her B-block class was 7 more than the number of C's she gave in her C block class. Everyone in her A- and B-block classes got an A or B. The median size of her three classes is 22 students. If the total number of A's she gave was one more than 7 times the number of C's she gave out, how many A's did she give in total?

6. (L4) Find the value of ♥ if

♥ + ☺ + ☼ = 13
☺ × ☼ + ☺ = 0
☼ × ♦ × ♦ = 9
☺ + ♦ = 1

Meet #3 Geometry: 2.1 Review of Meet 1: Angles in Polygons

1. (L1) The measure of an interior angle in a regular N-gon is 144°. What is N?

2. (L1) If ABCDE is a regular pentagon and ABF is an equilateral triangle, what is the measure of angle CBF?

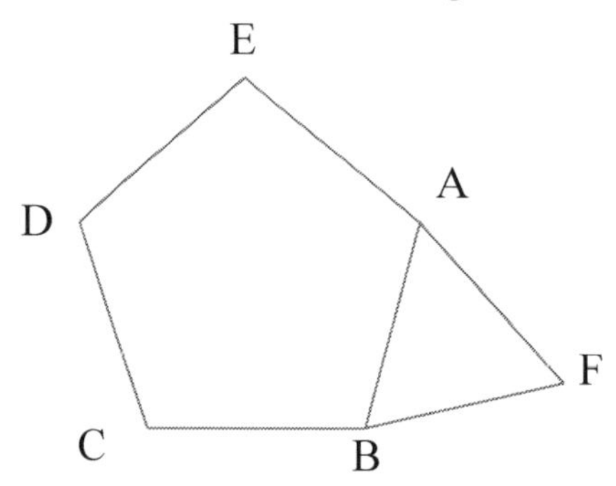

3. (L2) The interior angles of a regular N-gon and a regular M-gon are supplements. What is the largest possible value for N+M?

4. (L2) In the figure below CBF and EAF are straight lines and ABF is an isosceles triangle. What is the difference between the median and the mean of the measures of the angles of pentagon ABCDE.

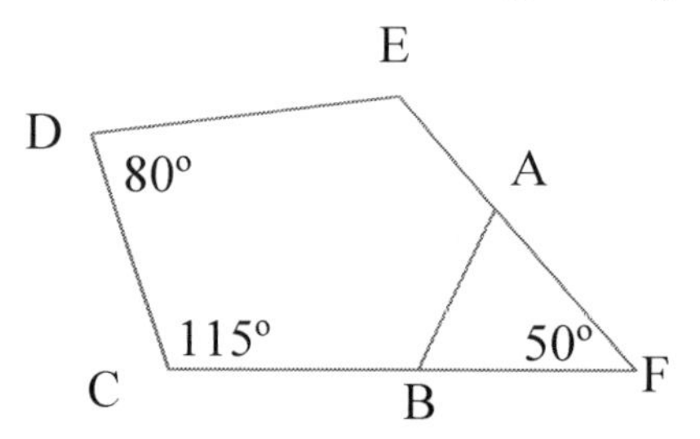

5. (L3) The measure of the interior angle in a regular N-gon is ten times the measure of the exterior angle in a regular M-gon. What is the largest possible value of M?

6. (L4) The measure of the interior angle in a regular N-gon is ten times the measure of the exterior angle in a regular M-gon. What is the sum of all of the possible values for N?

Meet #3 Geometry: 2.2 Review of Meet 2: Areas of Polygons

1. (L1) In trapezoid ABCD sides AB and DC are parallel. Angles ABC and BCD are both right angles. If AB=10cm, BC= 6cm, and CD=18cm, what is the area of ABCD in cm^2?

2. (L2) In the triangle on the left below AC = 13cm, BC=8cm, and the altitude AD divides BC in the ratio BD:DC = 3:5. What is the area of ABC?

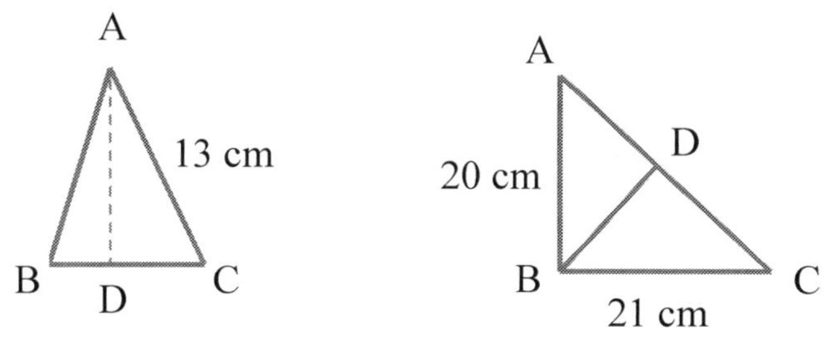

3. (L2) In the triangle shown on the right above ABC and BDA are both right angles. If AB=20cm and BC=21cm, what is the length of BD in cm?

4. (L3) In the triangle on the left below ABC is a right angle, BC=16cm, and the area of triangle ADC is $42cm^2$. What is the perimeter of ADC?

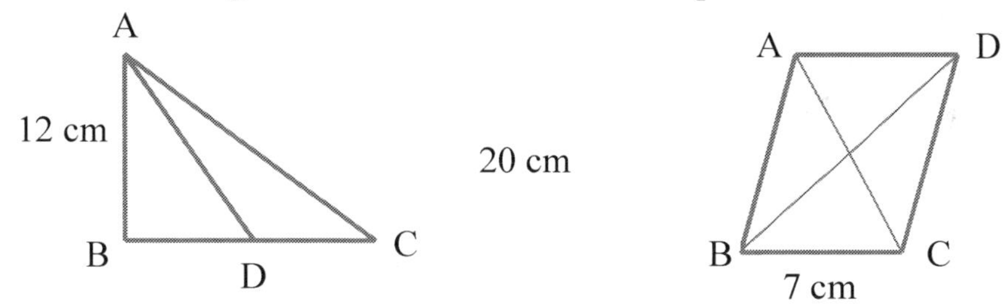

5. (L4) In parallelogram ABCD the length of BC is 7cm and the lengths of the diagonals are AC=13cm and BD=15cm. What is the area of ABCD?

6. (L4) In the figure below KEY, KAE, and ATY are all right angles and the length of EY is 63cm. What is the area of KATE?

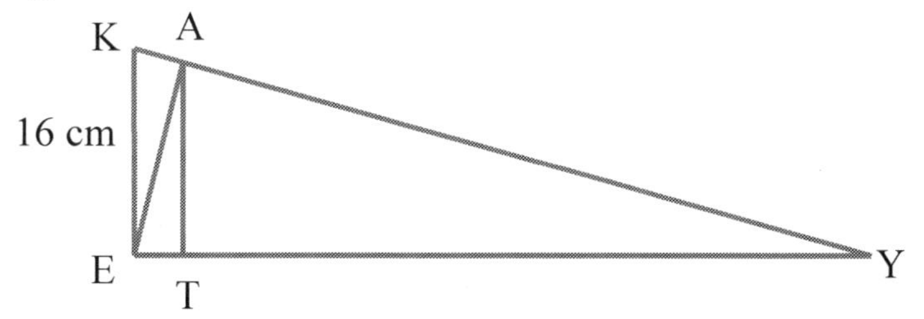

Meet #3 Geometry: 2.3 Diagonals in Polygons

1. (L1) How many diagonals can be drawn in a regular hexagon?

2. (L1) How many diagonals can be drawn in a convex 100-gon?

3. (L2) Avishek wanted to draw in all of the diagonals of the polygon below as dashed lines. He already drew in two of them. How many more does he need to draw?

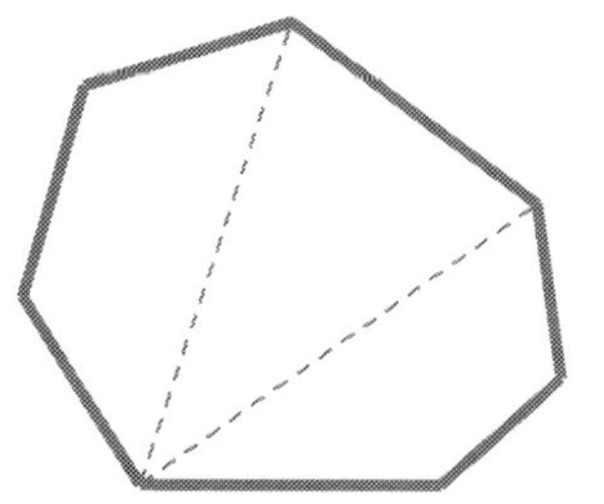

4. (L2) For what value of N is the number of diagonals in a regular N-gon closest to 100?

5. (L3) Find the sum of the values of N for which the number of diagonals in a regular N-gon is prime.

6. (L3) Roberto added up the number of diagonals in a convex N-gon for every N from 5 to 20. What number did he get?

7. (L4) Mei inscribed a regular N-gon and a regular M-gon in a circle. She noticed that the side length of the N-gon was 1 cm and that the M-gon had seven more diagonals than the N-gon? What is the smallest possible value for the area of the circle?

Meet #3 Geometry: 2.4 Pythagorean Theorem (part 1)

1. (L1) The legs of a right triangle are 6 cm and 8cm long. How long is the hypotenuse?

2. (L1) Two sides of a right triangle are 60 cm and 61 cm. How long is the third side if its length in centimeters is a whole number?

3. (L2) What is the area of quadrilateral ABCD if the length of the diagonal AC is $\sqrt{85}$ cm?

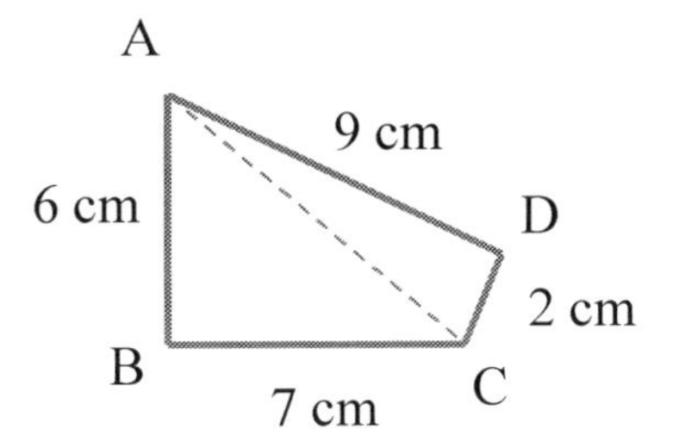

4. (L2) Find the area of triangle AFE if ABCD is a square with side length 4 cm, F is the midpoint of DC, and BE:EC = 3:1.

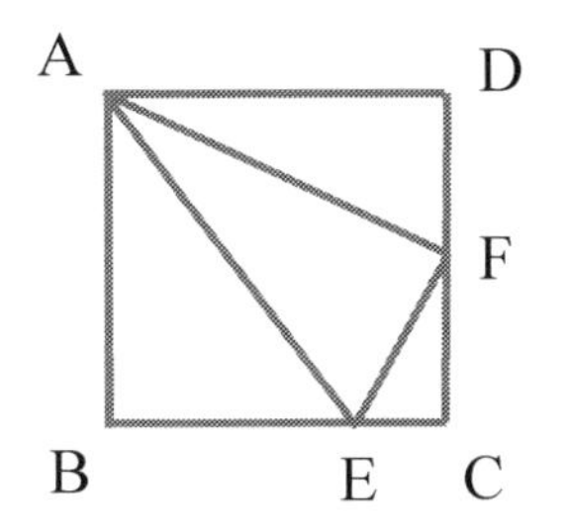

5. (L4) Pythagorean triple (a, b, c) is *primitive* if GCF(a, b, c) = 1. There are two primitive Pythagorean triples with $a < b < c$ and c=85: (13, 84, 85) and (36, 77, 85). A circle with radius 85 is centered at (0, 0). How many points with integer coordinates does the circle pass through?

Meet #3 Geometry: 2.4 Pythagorean Theorem (part 2)

6. (L1) What is the distance from the point (4, -1) to the point (7, 3)?

7. (L2) In the figure below angles ABC, ACD, and ADE are right angles. What is the length of AE?

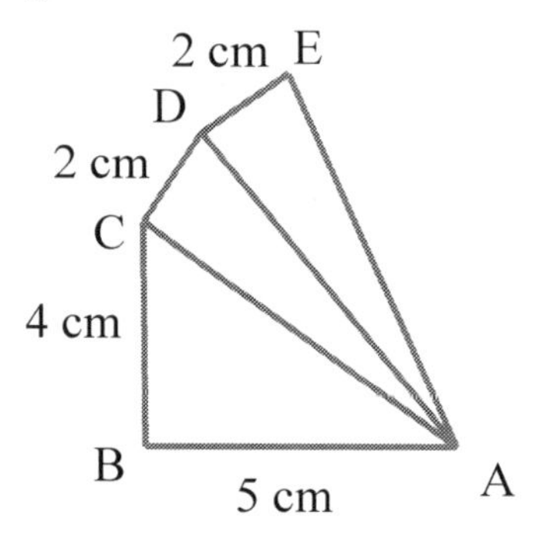

8. (L2) Helen's backyard is completely flat, but she was able to build a zipline by attaching a wire to a tree and to the crossbar of an old swingset. The top end of the wire is attached to tree that is directly in front of her back door 30 ft. away. She put a bolt in the tree 12 feet high and attached the wire to it. The crossbar of the swingset is 7 ft. off the ground. The wire attaches to it at a point that Helen reaches by taking five steps from her back door in the direction of the tree and then turning 90° to the left and taking 20 more steps. If her steps are exactly 2 ft., how long is the wire?

9. (L3) In nonconvex pentagon ABCDE vertices A, D, and C are collinear. AD is the perpendicular bisector of BE. The length of AE is 13cm, the length of BE is 24cm, the length of AD is 21cm, and length of AC is 40cm. What is the perimeter of ABCDE?

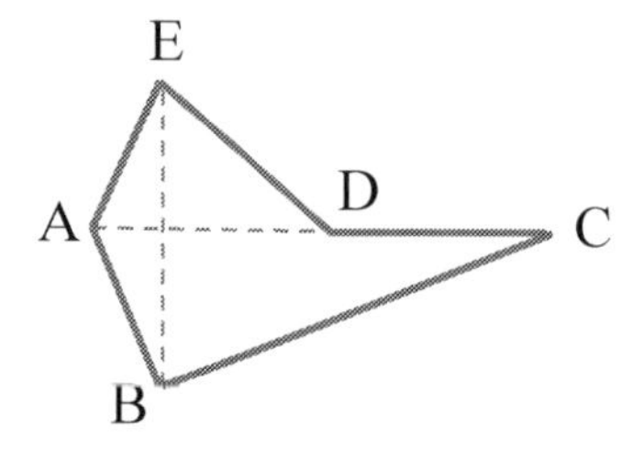

Meet #3 Geometry: 2.4 Pythagorean Theorem (part 3)

10. (L3) ABCD shown below is an isosceles trapezoid with bases AB and DC having lengths 2 in. and 5 in. respectively. The perimeter of ABCD is 16 in. What is the length of the diagonal DB as a decimal?

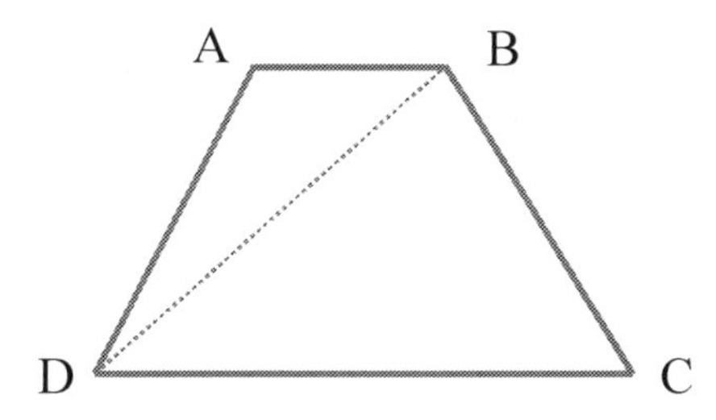

11. (L3) Gee Young lives in Manhattan where the streets have a regular grid pattern with rectangular blocks. There are 20 blocks per mile in the north-south direction and 4 blocks per mile going east-west. She leaves her apartment and walks 15 blocks north and 4 blocks west to a bakery. She buys a donut and then walks 2 blocks further west and 4 blocks further north to school. After school she walks 4 blocks east and 5 blocks north to a Starbucks with her friends. How many miles from her apartment is the Starbucks as the crow flies? Give your answer as a decimal.

12. (L4) What is the area in cm^2 of a trapezoid if its bases are 29 cm and 37 cm long and its other sides are 7cm and 9 cm long? Give your answer in simplest radical form.

13. (L4) ABC is an isosceles triangle with AB=AC=53 cm and BC=56 cm. Let E be a point on AC with EC=2 cm. Let D be a point on BC chosen so that DE is perpendicular to the line connecting A to the midpoint of DE. What is the length of BD in cm?

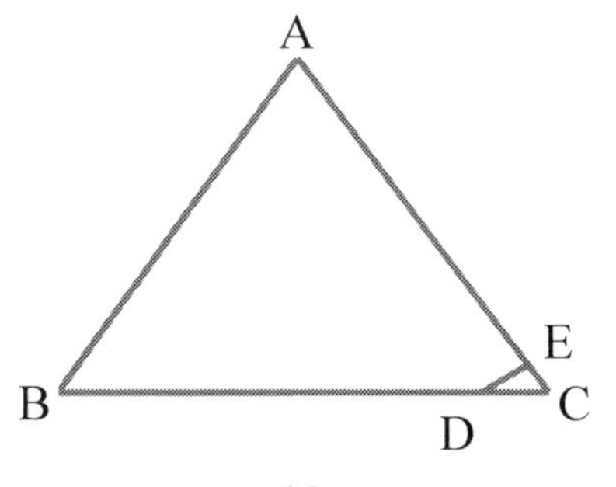

Meet #3 Number Theory: 3.1 Scientific Notation

1. (L1) Write each of the following numbers in scientific notation:

$54 =$ $328 =$

$163{,}000 =$ $0.04 =$

$0.0025 =$ $0.000123 =$

2. (L1) Write each of the following numbers in standard scientific notation:

$12.3 \times 10^{6} =$ $243 \times 10^{-5} =$

$800 \times 10^{-2} =$ $0.3 \times 10^{6} =$

$0.0024 \times 10^{-5} =$ $0.000008 \times 10^{1} =$

3. (L2) Find the products below and write them in scientific notation:

$(8 \times 10^{-2}) \times (3 \times 10^{6}) =$

$(1.6 \times 10^{4}) \times (3 \times 10^{6}) =$

$(1.4 \times 10^{-5}) \times (8.2 \times 10^{3}) =$

4. (L2) Write the answers to the problems below as decimal numbers.

$(8.2 \times 10^{-3}) \times (2 \times 10^{4}) =$

$(2.5 \times 10^{4}) \times (2.5 \times 10^{-6}) =$

5. (L2) Write the answers to division problems below in scientific notation:

$(8 \times 10^{5}) \div (2 \times 10^{3}) =$

$(2 \times 10^{4}) \div (5 \times 10^{6}) =$

$(3.6 \times 10^{-5}) \div (6 \times 10^{3}) =$

Meet #3 Number Theory: 3.2 IMLEM Scientific Notation Problems

1. (L1) Simplify the expression below. Write the result in scientific notation.

$$\frac{(8 \times 10^{5}) \times (6 \times 10^{-2})}{(2 \times 10^{3}) \times (3 \times 10^{-2})}$$

2. (L2) Simplify the expression below. Write the result in scientific notation.

$$\frac{(8.1 \times 10^{5}) \times (1.6 \times 10^{-4})}{(4 \times 10^{4}) \times (9 \times 10^{1})}$$

3. (L2) Simplify the expression below. Write the result in scientific notation.

$$\frac{(4.2 \times 10^{-5}) \times (1.08 \times 10^{-2})}{(2.7 \times 10^{3}) \times (1.2 \times 10^{-2}) \times (1.4 \times 10^{-11})}$$

4. (L2) Avogadro's number, N_0, is one of the most important numbers in chemistry. Its value is approximately 6.022×10^{23}. N_0 atoms of hydrogen weigh one gram. How much does one atom of hydrogen weigh in grams? Answer in scientific notation with one digit after the decimal point.

5. (L3) The distance from Boston to Seattle is about 2500 miles. Light travels at about 1.86×10^{5} miles per second. If you went back and forth between Boston and Seattle at the speed of light, how many round trips could you make in one hour? Give your answer in scientific notation with two digits after the decimal point.

6. (L3) Simplify the expression below. Write the result as a decimal.

$$\frac{(1.44 \times 10^{4}) \times (2.89 \times 10^{-2})}{(1.7 \times 10^{9}) \times (3.6 \times 10^{-2}) \div (1.25 \times 10^{3})}$$

Meet #3 Number Theory: 3.3 Basics of Bases

1. (L1) What is the base 10 value of the base 5 number 3043?

2. (L1) Write the base 10 number 55 in the base 8 system.

3. (L2) Write the base 2 number 10110 in the base 3 system.

4. (L2) What is the base 10 value of the base 16 number BB8?

5. (L2) Find the positive difference between the base 6 number 3333 and the base 4 number 3333. Express your answer as a base 10 number.

6. (L2) How many two-digit base 5 numbers are prime?

7. (L3) How many of the first 50 positive whole numbers have the property that the sum of their digits is the same whether they are written in base 10 or base 8?

8. (L4) How many factors does the base 5 number 131313 have?

Meet #3 Number Theory: 3.4 Converting from Base A to Base B

1. (L1) Write the base 8 number 63 as a base 5 number.

2. (L1) Write the base 4 number 312 as a base 2 number.

3. (L2) Express the binary number 101100101 as a base 8 number.

4. (L2) Express the base 9 number 6241 as a base 3 number.

5. (L3) Express the base 16 number A358 as a base 8 number.

6. (L3) What is the sum of the base 8 number 463 and the base 4 number 3020? Give your answer as a base 4 number.

Meet #3 Number Theory: 3.5 Arithmetic in Other Bases

1. (L1) Find the sum of the base 5 numbers 1234 and 310. Express your answer as a base 5 number.

2. (L1) What is the product of $175_{\text{base } 8}$ and $10_{\text{base } 8}$. Express your answer as a base 8 number.

3. (L1) Evaluate $111_{\text{base } 9} + 222_{\text{base } 9} + 333_{\text{base } 9} + 444_{\text{base } 9}$. Express your answer as a base 9 number.

4. (L2) Evaluate $2237_{\text{base } 8} - 140_{\text{base } 8}$. Express your answer as a base 10 number.

5. (L2) Find the product of $14_{\text{base } 6}$ and $34_{\text{base } 6}$. Express your answer as a base 6 number.

6. (L3) Find the product of $223_{\text{base } 8}$ and $1001_{\text{base } 2}$. Express your answer as a base 8 number.

7. (L3) For what value of b is the sum of the base b numbers $363_{\text{base } b}$ and $1032_{\text{base } b}$ equal to $1425_{\text{base } b}$.

8. (L4) Find N if $33_{\text{base } 8} + 37_{\text{base } 8} + 43_{\text{base } 8} + \ldots + N_{\text{base } 8} = 1260_{\text{base } 8}$. Give your answer as a base 8 number.

Meet #3 Number Theory: 3.6 Word Problems Related to Bases

1. (L1) A set of weights consists of "cubes" which are 5 cm on a side, "squares" which are 5 cm by 5 cm by 1 cm, "rods" which are 5 cm by 1 cm by 1 cm, and "minicubes" which are cubes 1 cm on a side. A minicube weighs 1 gram. How much does a peanut butter & fluff sandwich weigh in grams if it balances with 1 cube, 3 squares, 1 rod, and 2 minicubes?

2. (L1) Ms. Hall bought two dozen twelve-packs of water for the first math meet. If she had two full twelve-packs and two additional bottles left over, how many bottles were given out at the meet?

3. (L2) Emily used a 9 in. × 9 in. × 9 in. cube, a 9 in. × 9 in. × 1 in. square pan, and an ice-cube tray with twelve 1 in. × 1 in. × 1 in. spaces to measure the volume of a 20 pound bag of flour. She found that after she filled the cube once, she had enough flour left over to fill the square pan twice, then fill the ice-cube tray completely 3 times, and then finally fill four of the spaces in the ice-cube tray. How many cubic inches of flour was in the 20 pound bag?

4. (L3) For what value of b is the following statement true: Taylor Swift was born in the year $3705_{\text{base b}}$.

Meet #3 Number Theory: 3.7 Advanced Topic: Adding and Subtracting in Scientific Notation

1. (L1) Compute each of the sums below. Write the answers in scientific notation:

$(4.3 \times 10^6) + (3.1 \times 10^6) =$

$(8.3 \times 10^3) + (3.8 \times 10^3) =$

$(2.3 \times 10^4) + (1.1 \times 10^5) =$

2. (L2) Compute the sums and differences below. Write the answers in scientific notation:

$(4.44 \times 10^4) - (4.4 \times 10^4) =$

$(8.34 \times 10^3) - (3.8 \times 10^2) =$

$(4.3 \times 10^{-5}) + (7.1 \times 10^{-6}) =$

3. (L2) Find $(4.44 \times 10^{-3}) - (1.21 \times 10^{-2}) + (8.17 \times 10^{-3}) + 0.0101$. Give your answer in scientific notation.

4. (L3) Find $(2.6 \times 10^3) \times ((1.21 \times 10^{-5}) - (80 \times 10^{-7}))$. Give your answer as a decimal.

5. (L3) On the day of a solar eclipse the Earth was 1.496×10^{11} meters from the Sun. The moon was 1.49216×10^8 kilometers from the Sun. If a car drove 24 hours per day at 100 km/hour, how many days would it take to travel the distance from the Earth to the Moon on that day.

Meet #3 Arithmetic: 4.1 Basics of Exponents

1. (L1) Find $4^3 - 3^4$.

2. (L1) Compute $3^4 - 2 \times 3^2 + 3^0$.

3. (L2) Write 1.5^{-2} as a repeating decimal.

4. (L2) Find $4^{-1} - (\frac{1}{2})^2$.

5. (L3) What is the largest integer n for which $\left(\frac{3}{2}\right)^{-n} > (-2)^2 \times 40^{-1}$?

6. (L3) What whole number is closest to

$$5^1(1 + (-4)^{-1} + (-4)^{-2} + (-4)^{-3} + (-4)^{-4} + (-4)^{-5})$$

Meet #3 Arithmetic: 4.2 Operations with Exponents

1. (L1) Compute $2^5 \times 5^{-1} \times 2^2 \times 5^3$.

2. (L1) Write $\left(\frac{2}{3}\right)^{-2} + \left(\frac{3}{5}\right)^2$ as a mixed number.

3. (L2) Compute $(5^5)^2 \times 5^{-5} \times \left(\frac{1}{5}\right)^3$.

4. (L2) Write $\frac{5^{-3}(5^5+5^3)}{3^{-4}(3^6+3^4)}$ as a decimal.

5. (L3) Compute $\left(\frac{2}{3} \times \frac{4}{5} \times \frac{6}{7}\right)^2 \div \left(\frac{4}{3} \times \frac{6}{5} \times \frac{8}{7}\right)^2 \times (1.5^2 + 2^2)$.

6. (L4) Compute $\left(\left(\frac{5005}{1080}\right)^2 \left(\frac{777}{8181}\right)^{-3} \left(\frac{37}{429}\right)^3 \left(\frac{1030301}{1001}\right)^{-1}\right)^{-1}$.

Meet #3 Arithmetic: 4.3 Roots

1. (L1) Find $\sqrt{9} - \sqrt[3]{8}$. (In all questions assume all square roots are positive.)

2. (L1) Find $\sqrt{144} \div 16^{1/4}$.

3. (L2) Find $\sqrt[3]{7^6}$.

4. (L2) Find $(\sqrt{3} - \sqrt{2})(\sqrt{3} + \sqrt{2})$.

5. (L2) Write $\sqrt[4]{50^2}$ in simplest radical form.

6. (L3) Write $\sqrt[3]{4\sqrt{625}} \div \sqrt[6]{\left(\frac{2}{125}\right)^2}$ in simplest radical form.

7. (L3) Find $\sqrt{8 + 2\sqrt{15}} \times \sqrt{8 - 2\sqrt{15}}$.

8. (L3) What whole number is closest to the square root of the whole number that is closest to the cube root of 1700?

Meet #3 Arithmetic: 4.4 IMLEM Questions

1. (L2) Simplify $\left(\frac{2}{3}\right)^3\left(3\frac{1}{3}\right)^{-2}15^3$.

2. (L2) How many whole numbers are between $\sqrt[3]{10}$ and $\sqrt{103}$?

3. (L3) Simplify $(8^3 3^2 2^{-5})^2 \times \sqrt{7^0 + 2^3 3^1} \div \left(\frac{2}{3}\times\frac{3}{4}\times\frac{1}{6}\right)^{-4}$

4. (L3) Write $\sqrt[5]{3\sqrt{9^3 \times 3^5} \div \left(3^{-2} \times \sqrt[4]{3^{14}}\right)}$ in simplest radical form.

5. (L3) How many whole numbers are within $\sqrt{72}$ of $\sqrt{7^0 + 7^2}$?

6. (L4) Simplify the following expression:

$$\sqrt[2]{\left(\sqrt[3]{3^5 + 11^2 + \sqrt{222^2 \div \left(\frac{729}{64}\right)^{1/3}}}\right)^2 + \left(5\sqrt{13} + 10\right)\left(5\sqrt{13} - 10\right)}$$

7. (L4) What whole number is closest to $\frac{1}{\sqrt{117}-\sqrt{115}}$?

Meet #3 Algebra: 5.1 Linear Equations with Absolute Values

1. (L1) Find $|5 - |7 - 23| + 3|$.

2. (L1) What are the two possible values for x if $|x - 6| = 2$?

3. (L2) Find all possible values for x if $|2x - 7| = 3$.

4. (L2) Find the absolute value of the difference between the two solutions to the equation $|3x + 5| = 8$.

5. (L2) One half of the positive difference between a number and 7 is equal to 3. What is the smallest possible value for the number?

6. (L3) Find the sum of all possible values of x if $|7 - 2x| = x - 2$.

7. (L3) Find x if $|9 + 4x| = 6(x + 3) - x$.

8. (L4) Find all possible values of x if $\big|2 - |2 - x|\big| = |x - 3|$.

Meet #3 Algebra: 5.2 Working with Inequalities

1. (L1) For what value of A is does the solution to $x - 3 > A$ match the line graph shown below.

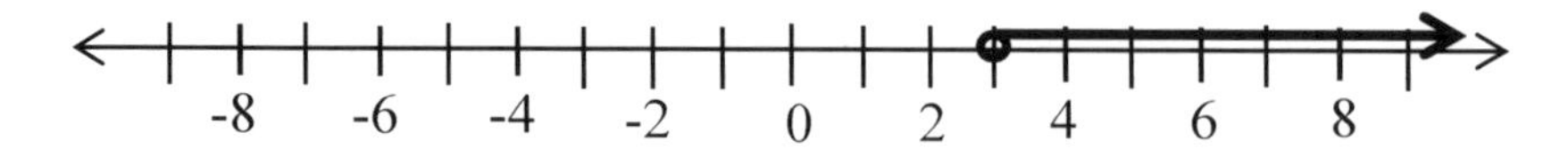

2. (L1) Solve for x: $3(x + 1) + 2 < x - 1$.

3. (L2) For what value of C is the line graph shown below the solution to the equation $3(x - 1) - Cx > 3$?

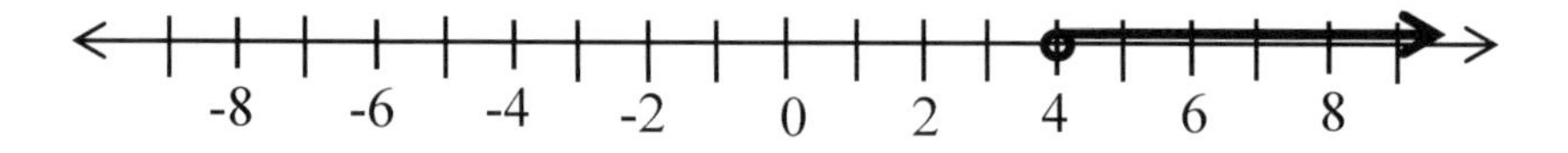

4. (L2) For what values of x is $11 + \frac{2}{x} < 17$?

5. (L3) For what value of A is $5(A - x) + 3(x + A - 1) \geq A$ true if and only if $x \leq 17$?

6. (L3) For what value of A is the line graph below the set of solutions to $337(x - A) + 7A(x + 1) - A(x - 335) \leq -7$?

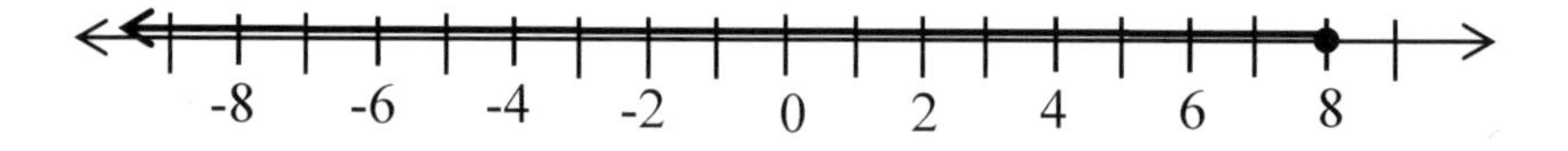

7. (L4) For what values of A is the set of all x with $|x| < 3$ the solution to $x(A - x) + A(x + A) > 2A(x + 4)$?

Meet #3 Algebra: 5.3 Working with Absolute Values and Inequalities

1. (L1) How many whole numbers n are solutions to $|n - 7| < 7$?

2. (L1) Find the sum of all integer solutions to $15/|n| \geq 3$?

3. (L2) How many integers n are not solutions to $\frac{11}{|n-1|} < 4$?

4. (L2) Find the median of the set of integer solutions to $|7x - 120| < 7$.

5. (L3) For what values of x is $|x - 3| < |x - 10|$?

6. (L3) What is the largest possible value for $|x|$ if x is a whole number and $\frac{15}{|3x+4|} > 2$?

7. (L4) Find the sum of all whole numbers that satisfy $|2x - 7| < |x + 7|$.

8. (L4) How many integers n satisfy $\frac{8}{|2n-3|} < \frac{2}{|n+{}^{1}/_{2}|}$?

Meet #4 Geometry: 2.1 Areas and Perimeters of Circles

1. (L1) The circumference of a circle is 8π centimeters. What is its area in square centimeters? Express your answer in terms of π.

2. (L1) Aimee rides her bike to school with her younger brother Evan. The tires on Evan's bike have a diameter of 12". Aimee's tires have a diameter of 24". If Evan's tires make 400 complete rotations on the way to school, how many rotations will Aimee's tires make?

3. (L2) The rectangle in the figure below has an area of 32 cm^2. What is the area in cm^2 of the region inside the rectangle and exterior to the circles?

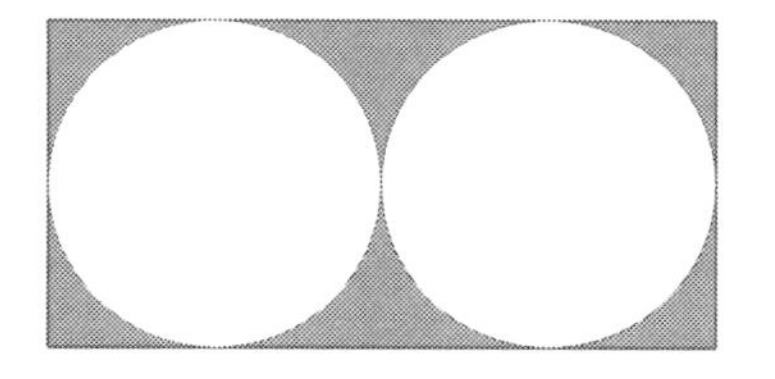

4. (L2) A circle is inscribed in a square of side length x cm. The area of the square in cm^2 is equal to the perimeter of the circle in mm? Find the side length of the square in millimeters rounded to the nearest whole number.

5. (L3) A square has an area of 49 square cm. One side of the square is a diameter of the circle on the right. A smaller circle is tangent both to the larger circle and to the midpoint of the opposite side of the square. What is the area of the shaded region in square cm? Give your answer as a decimal to the nearest tenth.

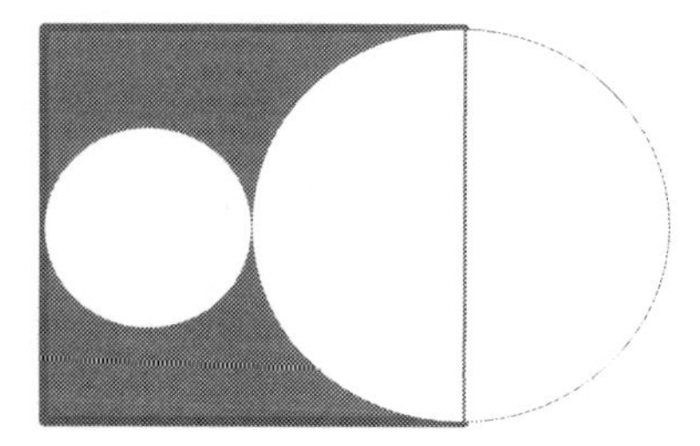

Meet #4 Geometry: 2.2 Arcs and Angles

1. (L1) The circle on the left below has its center at O and the measure of minor arc AB is 60 degrees. What is the measure of angle AOB?

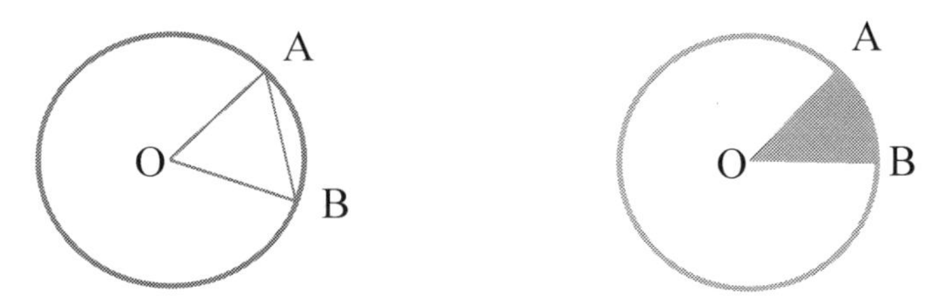

2. (L1) The circle on the right above the length of radius OA is 8cm and the measure of minor arc AB is 45°. What is the area of sector AOB in cm^2?

3. (L2) In the circle on the left below, O is the center of the circle, OAB is an equilateral triangle, and the measure of OAC is 20 degrees. What is the degree measure of angle CBO?

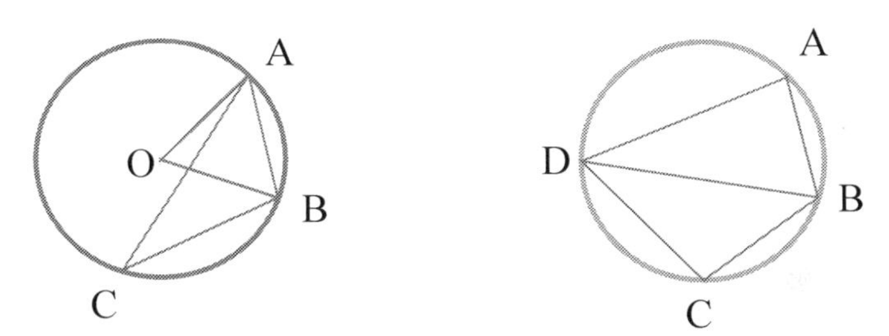

4. (L3) In the circle on the right above AB=BC and the measure of angle CDA is 40 degrees less than the measure of angle ABC. What is the degree measure of angle CDB?

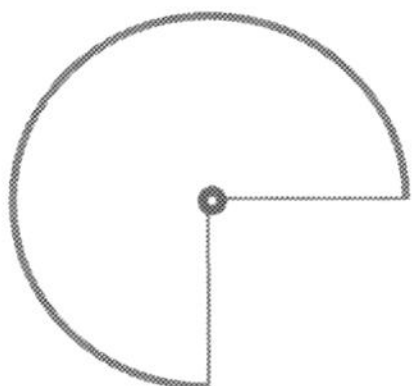

5. (L3) Vincent took a 24 inch diameter wooden disc and made it into a very bad wheel by using a saw to cut out one quarter of the disk as shown. If he starts with the wheel in the position shown and rolls it along the ground for a distance of 100 feet, how many complete revolutions will the wheel make? (Do not count the partial revolution at the end.)

Meet #4 Geometry: 2.3 A Quick Review of Triangles and Polygons

1. (L1) Regular octagon ABCDEFGH is inscribed in a circle or radius 6 cm. What is the measure in degrees of angle FEH?

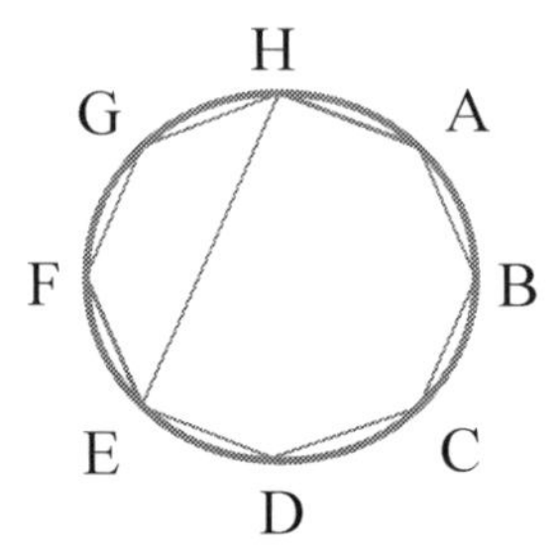

2. (L1) The sides of a triangle have lengths 4 cm, 5 cm, and 7 cm. The area of the triangle is $a\sqrt{6}$ cm^2. What is a?

3. (L2) The measures of the angles of a quadrilateral form an arithmetic progression. What is the measure of the largest angle if the measure of the second largest angle is 108º?

4. (L2) In triangle ABC the measure of angle A is 30º, the measure of angle C is 75º, $\overline{AD} = \overline{DB}$, $\overline{AE} = 6$ cm, and DE is perpendicular to AB. What is the length $\overline{AC}$?

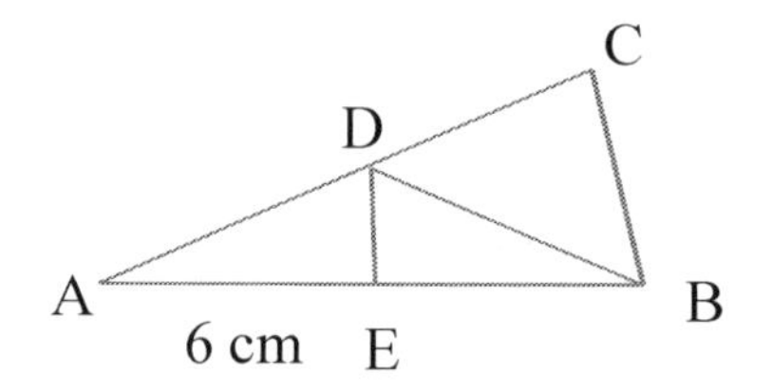

5. (L3) What is the area of a triangle with vertices (1, 1), (4, -1), and (3, 4)?

6. (L4) Regular hexagon ABCDEF is inscribed in a circle of radius 1 cm. Hexagon A′B′C′D′E′F′ is formed by connecting the midpoints of adjacent sides of ABCDEF. Hexagon A″B″C″D″E″F″ is formed by connecting the midpoints of adjacent sides of A′B′C′D′E′F′. What is the area of the region interior to ABCDEF and exterior to A″B″C″D″E″F″?

Meet #4 Geometry: 2.4 Advanced Circle Facts

1. (L1) In the figure on the left below the measure of minor arc AB is 60° and the measure of minor arc CD is 20°. What is the measure of angle AOB?

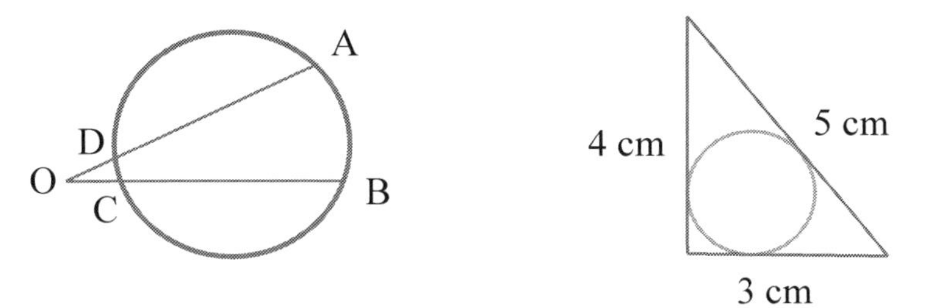

2. (L1) What is the radius in cm of a circle inscribed in a triangle with side lengths 3 cm, 4 cm, and 5 cm?

3. (L2) AB=9, AC=15, and BC=12. Find the ratio of the area of the ABC's circumscribed circle to the area of its inscribed circle as a common fraction.

4. (L2) A, B, C, and D are on a circle with AB=1, BC=2, CD=3, and DA=4. What is the area of quadrilateral ABCD in simplest radical form?

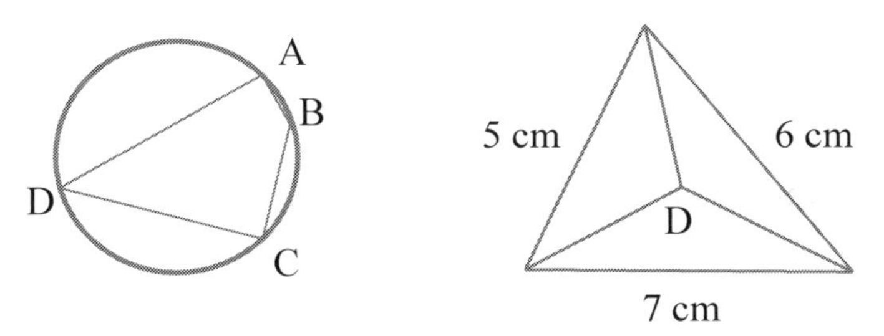

5. (L3) Let ABC be a triangle with side lengths 5 cm, 6 cm, and 7 cm. Let D be a point with DA=DB=DC. Find DA in simplest radical form.

6. (L3) Let ABC be an isosceles triangle with area 25 cm^2 and AC = 10 cm. Let ACDE be an isosceles trapezoid as shown with DE= 5 cm. If ABCDE is a cyclic pentagon, what is the measure of angle EBD?

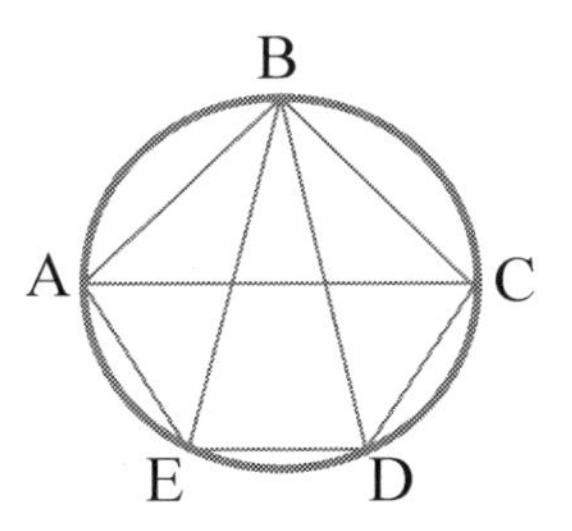

Meet #4 Number Theory: 3.1 Basic Sequences and Series

1. (L1) The first term of an arithmetic sequence is 13 and second term is 20. What is the 11th term?

2. (L1) What is the sum of the first 20 positive multiples of 5?

3. (L2) The nine numbers below form an arithmetic progression. What is their sum?

-29,517 -22,139 -14,761 -7,383 -5 7373 14,751 22,129 29,507

4. (L2) The 8th term of an arithmetic progression is 23. The 11th term is 224. What is the 26th term?

5. (L3) On January 1st, 2017 a zombie eats one brain. On January 2nd he eats two brains. He continues with this pattern throughout January eating 31 brains on the 31st. He then eats one brain on February 1st, two brains on February 2nd, and so on, finally eating 31 brains on December 31, 2017. How many brains in all did he eat in 2017?

6. (L3) The first two terms of an arithmetic sequence are 3.2 and 4.8. How many of the first 2017 terms are whole numbers?

7. (L4) A finite arithmetic sequence has common difference 7. One of its terms is 17. What is the smallest possible value for the sum of its terms if the sum is positive?

8. (L4) Consider an increasing arithmetic sequence of positive integers. One term is equal to 318. Another is equal to 303. None is equal to 9. What is the smallest possible value for the 20th term?

Meet #4 Number Theory: 3.2 Modular Arithmetic (part 1)

1. (L1) What time is it 275 hours after 3:30pm? Be sure to include am or pm in your answer.

2. (L1) A gym teacher divides kids into four teams by lining up the kids and assigning the first kid to team 1, the second to team 2, the third to team 3, the fourth to team 4, the fifth to team 1, the sixth to team 2 and so on. If there are 37 kids in the class what team is the last kid in line assigned to?

3. (L2) Simplify $13 + (5 \times 12)$ in mod 4 arithmetic.

4. (L2) What is the value of 3^{13} in modulo 7?

5. (L2) Bill can swim a lap of his school's pool in 45 seconds. He cannot swim 680 laps at this speed, but if he could and he starting swimming exactly at noon, what number would the minute hand on a traditional clock be pointing to when he finished?

6. (L2) Find all of the solutions to $3x + 3 = 7 \pmod{10}$ where x is chosen from $\{0, 1, 2, \ldots, 9\}$.

7. (L3) What is the smallest positive whole number x for which $2^{13}x - 1$ is a multiple of 17?

Meet #4 Number Theory: 3.2 Modular Arithmetic (part 2)

1. (L2) What is the value of $23(x-4)+2(x+45)$ in mod 12 arithmetic if $x=5$?

2. (L2) Find x if
   - x is a positive whole number.
   - x is less than 15.
   - $7x$ is five more than a multiple of 13.

3. (L3) What is the remainder when $5^{5555}+7^{7777}$ is divided by 4?

4. (L3) It takes Cecilia 23 minutes to walk around her block with her dog. She leaves her house at noon and decides to walk her dog around the block over and over again until she arrives back at her house at exactly 5 minutes after some hour. How many laps around the block does she end up walking?

5. (L4) What is the sum of the eleven smallest positive integers x for which $37(2x-1)$ ends with a 1 when written as a base 10 number?

6. (L4) For how many x in the set $\{0, 1, 2, \ldots, 100\}$ is $2^{2^x}+3^x$ a multiple of 5?

Meet #4 Number Theory: 3.3 Advanced Sequences and Series

1. (L1) What is the 10th term in the sequence below?

   1, 3, 6, 10, 15, …

2. (L1) A ball is dropped from the top of an 81 foot tall building. After it hits the ground it bounces up to a height of 54 feet. On each subsequent bounce it reaches two-thirds of the height from its previous bounce. How high does it reach after its fourth bounce?

3. (L2) A formula for the sum of the first n squares is

$$1^2 + 2^2 + \cdots + n^2 = \frac{n(n+1)(2n+1)}{6}$$

What is the sum of the first 10 even squares, i.e. what is $2^2 + 4^2 + \ldots + 20^2$?

4. (L2) Eleanor buys an iPhone for $649.99 on January 1, 2017. Every year its value decreases by 20%. On January 1st of year x Eleanor trades in her iPhone for a Samsung Galaxy 8 and finds that the value of her iPhone is less than $300. What is the smallest possible value of x?

5. (L3) What is the 100th term in the sequence below?

   3, 7, 13, 21, 31, 43, …

6. (L3) Seema draws a square of side length 1 cm. Then she puts a square of side length ½ cm next to it. Then she adds a square of side length ¼ cm, then a square of side length ⅛ cm, and so on. If she continues doing this forever, what will be the sum of the areas of all of the squares?

7. (L4) A ball drops from the top of a 100 ft. building. On its first bounce it bounces back to a height of 60 feet. On each subsequent bounce it bounces back to 60% of the height it reached on the previous bounce. What is the total distance in feet that the ball has travelled when it hits the ground for the 100th time? Give your answer as a decimal to the nearest tenth.

Meet #4 Arithmetic: 4.1 Percent Applications

1. (L1) What is 25% of 244?

2. (L1) The original price of pair of True Religion® jeans is \$89. How much would they cost if they were being sold at 30% off?

3. (L2) The Bigelow math team sold copies of *Hard Math for Middle School* for 16⅔% off of the list price. What percent greater than the price at which they sold the books was the list price?

4. (L2) What number is 37.5% greater than 50% of 176?

5. (L2) Sam's discount furniture's regular prices are 20% less than the MSRP. On President's Day they have a 10% off sale. What is the MSRP of a couch that they are selling for \$360 on President's Day?

6. (L3) Arvind wanted to buy an Xbox 720. Its regular price is \$449. If Arvind lives in a city with a 5% sales tax and has saved up \$420, what is the smallest whole number X for which Arvind can buy an Xbox 720 if it is being sold for at least X% off?

7. (L4) Kate's math class has 20 students. The number of students in Helen's class who got at least 90 on their last math test was 10% larger than the number of students in Kate's class who got at least 90. All scores were whole numbers and all students in both classes scored at least 80. The number of students in Helen's class is 25% larger than the number of students in Kate's class. The average score in Helen's class was X% larger than the average score in Kate's. What is the largest possible value for X?

Meet #4 Arithmetic: 4.2 Compound Interest

1. (L1) Stanford's tuition is $45,000. What will the tuition be two years from now if it goes up by 5% per year?

2. (L1) Yvonne's parents opened a college savings account with $1000 on September 1, 2005. If the account pays 4% interest compounded annually, how much money will be in the account on September 1, 2020?

3. (L2) A bank pays 3% annual interest compounded monthly. By what percent will the value of an account increase over an 8 year period if no deposits or withdrawals are made? Answer as a decimal to the nearest tenth.

4. (L2) A Certificate of Deposit paid a constant annual interest rate over a five year period. Interest was compounded annually. What was the annual interest rate if an investment of $25,000 grew to $32,000 over the five year period? Give your answer as a decimal to the nearest hundredth.

5. (L3) Martin has received an 8% raise in each of the past five years. Social security taxes are 7.65% of income for an individual with an income below $117,000. What was Martin's annual income 3 years ago if he paid $3854.72 in social security taxes this year? Round your answer to the nearest dollar.

6. (L4) On January 1st of each year a teacher's pension is increased by a cost-of-living adjustment which is one percentage point less than the previous year's increase in the Consumer Price Index (CPI). Suppose that the CPI increases by at least 1% in 2017, 2018, and 2019 and that the arithmetic mean of the increases in these three years is 2%. Given this information, what is the difference between the largest and smallest possible value for a teacher's pension in 2020 if the teacher receives a $30,000 pension in 2017?

Meet #4 Algebra: 5.1 Functions

1. (L1) Tastee Donuts charges $2.49 for a half dozen donuts, $3.49 for a dozen donuts, and 75 cents for a single donut. What would it cost to buy 15 donuts?

2. (L1) Kat's cell phone plan gives her unlimited calling and 500 texts a month for $40. Additional texts cost 1 cent. How much would she pay in a month if she made 434 minutes of phone calls and sent 656 texts?

3. (L2) Edison Electric charges its customers a $19.95 monthly fee, plus 13.4 cents per kilowatt hour for the first 50 kilowatt hours, and 14.4 cents per hour for all additional kilowatt hours. How much would they charge someone who used 160 kilowatt hours in a month?

4. (L2) People who are self-employed have to pay self-employment tax in addition to income taxes. Self-employment tax is 15.3% of the first $120,000 in income, and 2.9% of any additional income. If Josh had to pay $6120 in self-employment tax because of the money he earned tutoring Mathcounts, how much money did he make tutoring?

5. (L3) Desiree is offered a choice of two texting plans. Plan A has a fixed charge of $2.87 per month. It has 100 included texts. Plan B has a fixed charge of $7.87 per month. It includes 500 texts. Additional texts costs 2.2 cents under each plan. A sales tax of 6.25% is also applied to the bill under each plan. What is the smallest number of monthly texts for which Plan B is cheaper?

Meet #4 Algebra: 5.2 Two Equations in Two Unknowns

1. (L1) Find x if $x + 2y = 73$ and $x - y = 13$.

2. (L1) Genicia is 5 years older than her twin brothers. How old is Genicia if the sum of the three kids ages is 29?

3. (L2) The Fishers have chicken, cows, and dogs on their farm. One of their chickens is a rare mutant three-legged chicken. How many chickens do they have if their animals have a total of 37 heads and 111 legs?

4. (L3) Find B if $AB - A = 24$ and $2A + 4AB = 456$. Give your answer as a decimal.

5. (L4) Five years ago Anna's sister Hannah was three years older than Anna's younger sister Hana will be when Anna is in 8th grade. The sum of Anna's grade and the average of Hana and Hannah's ages is thirteen more than one-third of the sum of Hana and Hannah's ages. Three years from now the difference between half of Hana's age and Anna's grade will be 5. How old is Hannah?

6. (L4) A trapezoid has two right angles. The lengths of its four sides form an arithmetic progression, with one of the bases being the shortest side. What is the area of the trapezoid in cm^2 if its perimeter is 12cm?

Meet #4 Algebra: 5.3 N Equations in N Unknowns

1. (L1) What is the value of z if (x, y, z) is a solution to

$$\begin{aligned} x + 2y + z &= 32 \\ 2x - y &= 3 \\ y &= 7 \end{aligned}$$

2. (L1) The sum of Anna's age and Beryl's age is 20. The sum of Beryl's age and Claudio's age is 40. The sum of Anna's age and Claudio's age is 34. What is the sum of Anna's, Beryl's, and Claudio's ages?

3. (L2) On a recent math test Rhianne scored 23 points higher than Ryan. Rian had scored 15 points higher than Ryan. The average of Rhianne's Rian's scores was 87. What score did Ryan get on the test?

4. (L2) Angél, Annika, Mark, Melissa go to McDonalds. Annika pays $1.72 for a cheeseburger and a small fries. Mark gets 2 cheeseburgers, a small fries, and a large coke for $4.20. Melissa pays $2.22 for a small fries and a large coke. How much did Angél pay if he got a cheeseburger, two small fries, and a large coke?

5. (L3) A car company is considering using three different aluminum alloys. One is 95% aluminum and 5% magnesium. The cost of the metals needed to make it would be 88 cents/pound. Another possible alloy is 92.5% aluminum, 5% zinc, and 2.5% magnesium. It would cost 85 cents/pound. A final possibility would be 90% aluminum, 5% magnesium, and 5% zinc. It would cost 89 cents per pound. What is the cost per pound of aluminum?

6. (L3) A school's team score on the AMC 12 is the sum of the scores of its three highest scoring students. The sum of Oldtown South High School's top four students was 444. The difference between Anna's score and Lib's score was 24. Anna's score as was nine points better than the average of Gniy's and Hsivak's scores. If the four students mentioned above were the top four scorers at Oldtown South HS and all but one of them scored at least 100, what was their team score?

Meet #4 Algebra: 5.4 Time-Distance-Speed Problems

1. (L1) John walks to school at a speed of 3 miles per hour. How many minutes does it take him to get to school if school is ¼ mile from his house?

2. (L1) In problems in which someone drives an **equal distance** at speeds of r and s, their average speed for the whole trip is the harmonic mean of r and s. The harmonic mean of a and b is defined to be $\frac{2ab}{a+b}$. What is the harmonic mean of 15 and 45? Give your answer as a decimal.

3. (L2) It would take Yotam 4 hours to paint a room. Kamal is a faster painter and could do the job in 2 hours. If Yotam starts painting at noon, Kamal joins him at 1pm, and they then work together until the job is done, at what time do Yotam and Kamal finish?

4. (L3) Tom runs for 15 minutes at a speed of 10 miles per hour. At this point he is very tired so he jogs the next mile at a 12 minute-per-mile pace. This makes him feel better so he finishes by running 5 more minutes at 10 miles per hour. What is his average speed in miles-per-hour for the full run? Give your answer as a decimal.

5. (L3) Robin usually drives to school at 30 miles per hour. When it snows she only drives at 20 miles per hour. If it takes her five minutes longer to get to work when it snows, how far is her drive to work?

6. (L4) Tom bikes around a lake at a constant speed. If he had gone 5 km/h faster he would have finished two hours earlier. If he had gone 10 km/h faster he would have finished three hours earlier. How far did he bike in km?

Meet #5 Geometry: 2.1 Surface Areas and Volumes

1. (L1) The interior dimensions of a sandbox are 6 feet by 6 feet by nine inches. How many cubic feet of sand would be needed to fill it?

2. (L1) Justin's bedroom is 9 feet long and 12 feet wide. The ceiling is 7 feet high. What is the total surface area of the four walls in square feet?

3. (L2) Arvind wants to build a 20 foot long, four foot high, and 8 inch thick wall using rectangular stone blocks that are 4 inches by 6 inches by 8 inches. How many blocks will he need?

4. (L2) A red cube has a surface area of 726 cm^2. A blue cube has a volume that is one-eighth as large as the volume of the red cube. What is the surface area in cm^2 of the blue cube?

5. (L3) Ellie is planning to make a tent out of canvas. She plans to make it in the shape of a triangular prism. The bottom of a tent will be an 8 foot by 6 foot rectangle. The front and back will be triangles that are 6 feet wide and 4 feet high. How many square feet of canvas will she need?

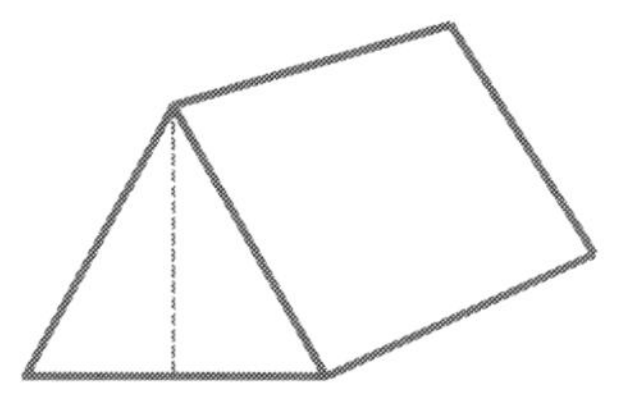

6. (L3) Lucine fills a measuring cup with exactly 1 liter of water. She drops three identical metal cubes in to the cup and the water rises to a level of 1 and three-eighths liters. What is the side length of the cubes in cm? (Recall that one liter is 1000 cubic centimeters.)

Meet #5 Geometry: 2.2 Shapes Related to Circles

1. (L1) A small ball has a surface area of 36π square inches. A larger ball has a radius twice as large. What is the surface area of the larger ball?

2. (L1) A right circular cone has a base diameter of 10cm and a height of 12cm. What is its surface area?

3. (L2) At Kate's Ice Cream they make a single-scoop cone by filling the cone completely with ice cream and adding a half ball of ice cream on top. If the cone has a diameter of 6cm and is 12cm high, what is the total volume in cm^3 of a single scoop cone? (Give you answer in terms of π.)

4. (L2) A silo is shaped like a cylinder with a half sphere on top. The cylindrical part is 20 feet in diameter. The total height is 60 feet. If one gallon of paint covers up to 350 square feet, how many gallons will the owner need to buy to paint his silo? (Assume paint is sold only in one gallon cans.)

5. (L3) Tennis balls have a diameter of 2.64 inches. They are traditionally sold with three balls stacked on top of each other in a cylindrical can. If the can is as small as possible without compressing the balls, what fraction of the volume of the can is empty space outside the balls? Give your answer as a decimal to the nearest hundredth.

6. (L4) Vibha is building a model volcano. She starts by making a right circular cone with a base diameter of 6cm and a height of 4cm. She then drills a hole with diameter 1cm, starting at the apex and going down through the center of the base. What is the surface area after this step?

Meet #5 Geometry: 2.3 Prisms and Pyramids

1. (L1) The Great Pyramid at Giza was built as a perfect rectangular pyramid. The base was a square with side length 230m. Its height was 150m. What was its volume in cubic meters?

2. (L1) A right rectangular prism has sides of length 8cm and 10cm and a volume of 720cm^3. What is its surface area in cm^2?

3.(L2) Two of the sides of a right square prism have lengths of 3cm and 5cm. What is the positive difference between the largest and smallest possible values for its volume in cm^3?

4.(L2) The base of a rectangular pyramid is a 6cm by 8cm rectangle. The four edges connecting the corners of the base to the top of the pyramid have length 13cm. What is the volume of the pyramid in cm^3?

5. (L3) The base of a square pyramid has a side length of 10 in. If the pyramid has a surface area of 240 square inches, what is its volume in cubic inches? Give your answer as a fraction in simplest radical form.

6. (L3) The base of a hexagonal pyramid is a regular hexagon with side length 10cm. The other six edges of the pyramid have length 20cm. What is the volume of the pyramid in cm^3?

Meet #5 Geometry: 2.4 Polyhedra

1. (L1) What is the positive difference between the number of edges of a cube and the number if vertices of a cube?

2. (L1) A regular octahedron has 8 triangular sides. How many edges does it have?

3. (L2) A dodecahedron has 20 vertices and 12 pentagonal faces. What is the positive difference between the number of space diagonals and the number of surface diagonals of a dodecahedron?

4. (L3) Amalie designs an alternative shape for a soccer ball. Her shape is obtained by cutting off each of the corners of a regular dodecahedron. Specifically, she puts dots at the midpoints of the three edges that connect to each vertex and cuts off the corner along the plane containing the three dots. How many vertices, edges, and faces does her shape have?

5. (L3) The surface area of a right rectangular prism in cm^2 is equal to its volume in cm^3. If two edges of the prism have lengths of 4cm and 8cm, what is the length in cm of its space diagonals?

6. (L3) Artur has three toothpicks of length 2 cm and three toothpicks that are $2\sqrt{2}$ cm long. He makes a triangular pyramid by first making a right triangle using two 2cm-long and one $2\sqrt{2}$ cm-long toothpicks. He uses this triangle as the base. He connects the final 2cm-long toothpick to the right angle of this triangle, making it perpendicular to the plane of the triangle. He connects the other two vertices of the base to the top of this toothpick using the other two long toothpicks. What is the volume of his pyramid in cm^3?

Meet #5 Geometry: 2.5 Units of Measurement

1. (L1) A rectangular field is 99 feet long and 50 feet wide. What is its area in square yards?

2. (L1) The scale on a map is 1 inch = 2.5 miles. What is the total distance traveled in miles if Josh drives east along a segment of a road that is 3 inches long on the map, then goes north along a segment of a road that is 4 inches long on the map, and then drives directly home along a straight diagonal road?

3. (L2) A cone has a volume of 14,875 cubic centimeters. What is its volume in cubic meters? Give your answer as a decimal to the nearest thousandth.

4. (L2) A marathon course is 26 miles and 385 yards long. If 32,000 people run the Boston marathon, what is the sum of the distances run by all of the people who run the marathon in miles? (A mile is 5280 feet long.)

5. (L3) The volume of a cube in cubic centimeters is numerically equal to its suface area in square millimeters. What is the volume of the cube in cubic yards? Given your answer to the nearest whole number. (Use 39.36 inches for the length of one meter.)

6. (L3) A right circular cone has a height of 10cm and a volume of one liter. If the cone is shortened to a height of 9cm by cutting off a slice with a thickness of 1cm parallel to the base, what is the volume of the shortened cone in cubic millimeters?

Meet #5 Geometry: 2.6 Advanced Topics

1. (L2) A cone has a height of 9cm and its base is a circle with radius 6cm. What is the volume of the frustum obtained by cutting the cone along a plane parallel to its base one-third of the way from the base to the apex?

2. (L2) Assume (incorrectly) that the Earth is exactly a sphere with a radius of 3950 miles. Suppose that three points on the surface of the Earth form an equilateral triangle and are far enough apart so that each angle measures exactly 61°. What is the area in square miles of the spherical triangle that they make? Give your answer to the nearest whole number.

3. (L3) Jack first cuts a 12 inch diameter globe along the equator, and then cuts the northern hemisphere along the 10° W and 70°W longitude lines. What is the surface area in cubic inches of the smallest piece that he makes?

4. (L3) A sphere with volume of 288π cm^3 is sitting on a table. It is cut into three pieces by making cuts parallel to the surface of the table one-quarter and three quarters of the way from the bottom to the top. What is sum of the surface areas of the three pieces that are formed?

5. (L3) A model used to show the Earth's layers was made by cutting away almost all of one quarter of a sphere with diameter 30cm. The cuts were made along the 0° and 90° W longitude lines. They connect the north and south poles and extend all the way to the center. One quarter of a sphere with radius 3cm was then glued back into the center to show the Earth's solid core. What is the surface area of the model?

Meet #5 Number Theory: 3.1 Venn Diagrams

1. (L1) In the Venn Diagram below what is the sum of the elements of $A \cap B$?

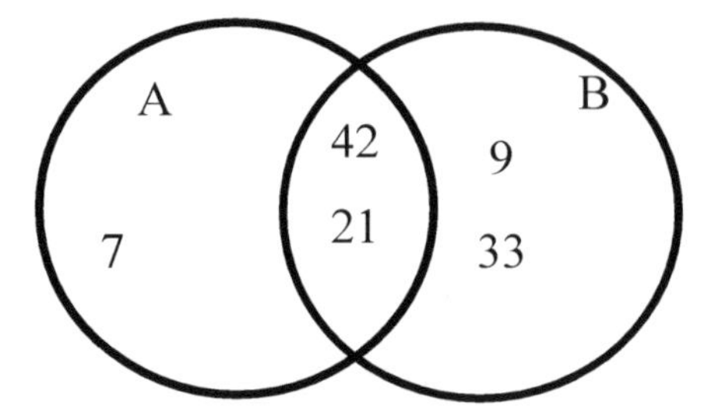

2. (L2) Only 17 of the 30 teams in the NFL have ever won the Super Bowl. Thirteen different teams have had the first overall draft pick since the year 2000. Six of them are teams that have won the Super Bowl. How many NFL teams have not won a Super Bowl and also not had the first overall draft pick since 2000?

3. (L2) The most prestigious high school contest for students in grades 10 and below is the USAJMO. You can't just decide to take it – you need to do well enough on the AMC 10A or AMC 10B to qualify to take the AIME. And you need to do well on the AIME (and the AMC 10) to get invited to take the USAJMO. In 2013 222 students were invited to take the USAJMO. 207 of them had taken the AMC 10A and 176 had taken the AMC 10B. How many of them took both tests?

4. (L3) Vivaan made a Venn Diagram containing the names of the members of his math class. Set A was the set of students who like Venn Diagrams, set B was the set of students who know how to draw a five-set Venn Diagram, and set C was the set of students who know how to spell "Venn". Eleven students in his class like Venn Diagrams. Nine of them know how to spell "Venn". Vivaan is the only student in his class who knows how to draw a five-set Venn Diagram. He likes Venn Diagrams, but misspells "Venn". Eight other students in the class also misspell "Venn". If there are 20 students in the class in total, how many students in the class are in exactly one of A, B, and C?

Meet #5 Number Theory: 3.2 Basic Set Theory

1. (L1) In the Venn diagram below, what is $|A \cup B| - |A \cap B|$?

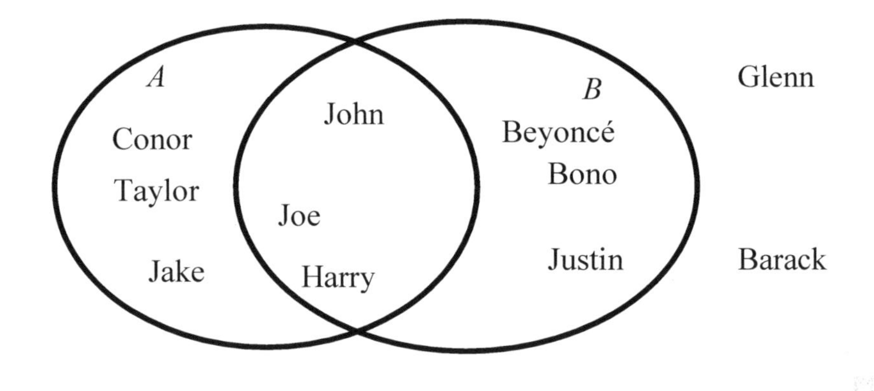

2. (L2) How many elements are in the set $A \cup (B \cap C)$ if A={E, G, L, N}, B = {A, C, E, I, L, N, O, R}, and C = {E, I, L, N, O, S}?

3. (L2) Find $(A \cup B) \cap (C \cap D)$ if $A = \{1, 3, 5, 7, 9, 11, 13, 15\}$, $B = \{5, 10, 15, 20, 25, 30\}$, $C = \{1, 2, 3, 5, 8, 13, 21\}$, and $D = \{2, 3, 5, 7, 11, 13, 17\}$.

4. (L3) Let A be the set of numbers that are 3 more than a multiple of 5. Let B be the set of all prime numbers. Let C be the set of two-digit numbers. Find $((A \cup B) \cap C) \cap (A \cap B)$.

5. (L3) Let P be the set of all palindrome numbers (numbers that remain the same if you reverse their digits). Let Z be the set of positive integers with a 0 in the tens place. Let T be the set of three digit positive integers. Let E be the set of numbers that are multiples of 11. Let S be the set of perfect squares. What is the sum of the elements of $((P \cup Z) \cap S) \cap (E \cap T)$?

Meet #5 Number Theory: 3.3 Inclusion-Exclusion Counting

1. (L1) Seven of Kyle's friends are in his English class and five are in his math class. How many of his friends are in at least one of the two classes if he has only one friend who is in both classes?

2. (L2) How many two-digit counting numbers contain the digit 7?

3. (L2) How many numbers from 1 to 1000 are multiples of 27 or 37?

4. (L3) How many positive three digit integers have at least two digits that are the same?

5. (L3) A positive integer is "seveny" if it has 7 as a digit or is a multiple of seven. How many positive numbers less than 100 are seveny?

6. (L3) There are 112 students at a math contest. 42 have taken the AOPS Intro to Algebra A course. 34 have taken AOPS Intro to Number Theory. And 38 have taken AOPS Intro to Counting & Probability. 21 students have taken exactly two of the three courses. And 11 have taken all three. How many students at the contest have taken at least one AOPS course?

7. (L3) How many positive three digit multiples of 3 have at least two 3's as digits?

8. (L3) How many of the first 1001 counting numbers are divisible by exactly two of 7, 11, and 13?

Meet #5 Number Theory: 3.4 Review of Modular Arithmetic

1. (L1) Let A be the set of natural numbers that are 2 more than a multiple of 5. Let B the set of natural numbers less than 1000. Find $|A \cap B|$.

2. (L1) Find the sum of the two smallest positive integers that are 3 more than a multiple of 5 and 5 more than a multiple of 7.

3. (L2) Let A be the set of natural numbers that are 2 more than a multiple of 5. Let B the set of natural numbers that are 1 less than a multiple of 17. Let C be the set of natural numbers that are divisible by 18 and 34. What is the smallest element of $(A \cap B) \cup C$?

4. (L2) Let A be the set of natural numbers that are 3 more than a multiple of 5. Let B be the set of natural numbers that are 3 more than a multiple of 13. Let C the set of natural numbers less than 1000. Find $|(A \cap B) \cap C|$.

5. (L3) Let A be the set of natural numbers that are 5 more than a multiple of 10. Let B be the set of natural numbers that are 10 more than a multiple of 25. Let C be the set of multiples of 17. Let D be the set of numbers less than 1000. How many elements are in the set $(A \cap B) \cap (C \cap D)$?

6. (L3) How many ordered pairs (x, y) with $0 \leq x \leq y \leq 100$ satisfy $x + y = 3\ (mod\ 8)$ and $y = 5\ (mod\ 22)$.

Meet #5 Number Theory: 3.5 Advanced Topic: Venn Diagrams and Inclusion-Exclusion with Four or More Sets

1. (L2) How many of the counting numbers between 100 and 200 are divisible by at least one of 11, 13, 17, and 19?

2. (L3) Maya came up with the clever diagram below to prove that it isn't impossible to draw a 4-set Venn Diagram if you can only draw circles – she used three standard circles for the sets *A*, *B*, and *C*, and then made set *D* the area between the two dashed circles.
In this diagram let *A* be the set of two digit multiples of 7. Let *B* be the set of two digit multiples of 9. Let *C* be the set of two digit multiples of 11. Let *D* be the set of perfect squares. Let x, y, and z be the number of elements in the parts of the diagram labeled below. What is $x + y + z$?

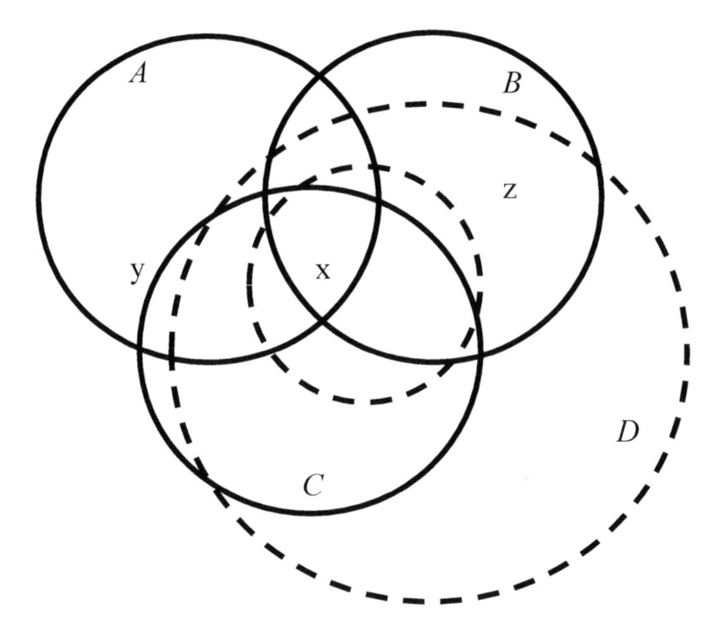

3. (L4) Six balls numbered 1 through 6 are to be placed in four bags labeled A, B, C, and D. In how many ways can this be done if at least one ball must be placed in each bag?

Meet #5 Arithmetic: 4.1 Combinatorics (part 1)

1. (L1) Anne, Anna, Ana, Anabel, and Anibal are giving presentations in Ms. Liang's class. In how many different orders can they do this?

2. (L1) Max's Overly Healthy Pizza offers eight possible toppings: Kale, Arugula, Broccoli, Onions, Fennel, Spinach, Artichokes, and Green Peppers. How many possible two-topping pizzas can one order there?

3. (L2) 101 dalmatians are entered in a dog show that names one dog the first prize winner, one as the second prize winner, and one as the third prize winner. How many different outcomes are possible?

4. (L3) Ms. Carrier wants need to choose three students from her math team to help her carry snacks in from her car. She notices that there are 220 different ways in which she could choose the set of three students. How many students are on the math team?

5. (L3) How seven letter sequences have a double L and can be rearranged to spell the name ELLISON?

6. (L3) A math team with six students is going to a math meet by car. The coach can take up to four students in his car. And Kevin's mother can take up to four students in her car. In how many different ways can the students be assigned to the two cars if Kevin must ride in his mother's car?

Meet #5 Arithmetic: 4.1 Combinatorics (part 2)

1. (L1) A coin in flipped seven times. How many different sequences of Heads and Tails are possible? (To be clear, HHHHHHT and HHHTHHH should be counted as two different sequences.)

2. (L1) A pizza shop offers eight toppings. How many different pizzas can you order there if you can order a plain pizza or add any number of toppings up to eight? (Each topping can only be added once, e.g. you cannot order double pepperoni.)

3. (L2) A set with N elements has fewer than 1,000,000 subsets. What is the largest possible value of N?

4. (L3) An ice cream shop has 36 flavors. A teacher tells her students that they can order one or two scoops (which can be two of the same flavor or two different flavors) in a cone or cup. If they get it in a cone they can have it dipped in chocolate. How many different orders can each student make? (Students cannot specify which flavor is on top in a two-scoop cone.)

5. (L4) A father buys food for his three children at McDonalds. He tells each that they need to choose exactly one of the five available sandwiches, they can choose to have a small, a large, or no French fries, and that they must either order milk or a small serving of any one of the five available sodas. The father adds up all the requests and makes a single order at the counter, e.g. "I'd like two Big Mac's, a McChicken sandwich, two large fries, and three milks." How many different orders might the father make?

Meet #5 Arithmetic: 4.2 Basic Probability

1. (L1) A six-sided die is rolled twice. What is the probability that the two numbers add up to five?

2. (L1) Two twenty-sided dice are rolled. What is the probability that the two numbers are the same?

3. (L2) Two six-sided dice are rolled. What is the probability that the product of the two numbers is a perfect square?

4. (L2) A fair coin is flipped three times. What is the probability that Heads does not occur twice in a row?

5. (L3) Two six-sided dice are rolled. What is the probability that the largest number is prime?

6. (L3) Balls numbered 1 through 10 are placed in a bag. Two distinct balls are drawn out at random. What is the probability that the difference between the larger and the smaller number is at least two?

7. (L4) Bill rolls two twenty-sided dice. What is the probability that the product of the two numbers is a perfect nth power for some positive integer $n>1$?

Meet #5 Arithmetic: 4.3 Probability Problems with Two Events

1. (L1) Two standard six-sided dice are rolled. What is the probability that both numbers are multiples of 3?

2. (L1) A bag contains 5 red marbles, 4 blue marbles and 3 green marbles. Amy takes out one marble and holds it in her hand. Blanche then takes out a second marble. What is the probability that Amy's marble is green and Blanche's is blue?

3. (L2) Balls numbered 1 to 10 are placed in a bag. One ball is drawn out at random. It is then placed back in the bag and a second ball (which could be the same as the first) is drawn out at random. What is the probability that at least one of the balls drawn has a prime number on it?

4. (L2) Balls numbered 1 to 10 are placed in a bag. Three distinct balls are drawn out at random. What is the probability that all three numbers are even?

5. (L3) Three twenty-sided dice are rolled. What is the probability that the product of the three numbers is even?

6. (L3) An IMLEM team has five boys and five girls. If the kids line up in a random order what is the probability that no two girls are next to each other?

Meet #5 Arithmetic: 4.4 Using Conditional Probabilities for Complicated Events

1. (L1) Anna flips a fair coin that has a 1 on one side and a 2 on the other. Bernice rolls a six-sided die. What is the probability that the number on the top of Anna's coin is a factor of the number on the top of Bernice's die?

2. (L2) Mathcounts® tests sometime ask directly about conditional probabilities. One way to think about conditional probability of E given X is to write out the complete set of all possible outcomes, cross out outcomes that do not satisfy X, circle the outcomes that satisfy E, and then count the fraction of the non-crossed out outcomes that are circled. Ben used this method to find the probability that the sum of the numbers on two dice is equal to 10 conditional on at least one number being 5. How many outcomes are in the grid he writes down? How many does he cross out? How many non-crossed out outcomes does he circle? What is the answer?

3. (L3) Use the method of question 2 to find the probability that the sum of the numbers on two dice is prime conditional on the sum being less than ten.

4. (L3) Two eight-sided dice are rolled. What is the probability that the product of the two numbers is a multiple of eight?

5. (L3) Balls numbered one to 100 are put in a bag. Two distinct balls are drawn out at random. What is the probability that the product ends in a one?

6. (L3) A two-element subset of {1, 2, …, 102} is chosen at random. What is the probability that the sum of the numbers in the subset is a multiple of 5?

7. (L4) A fair coin is flipped seven times. What is the probability that Heads comes up four times in a row at some point?

Meet #5 Arithmetic: 4.5 Averages

1. (L1) A math class has 14 girls and 6 boys. What is the class average on a test if the girls average 90 and the boys average 80?

2. (L2) Rebecca took 12 free throws in the first half of a game and made 75%. In the second half of the game she made all of her free throws. How many free throws did she attempt in the second half if she made 80% of her total attempts for the game?

3. (L2) Chen's math class has 6 tests per term. Scores on the tests range from 0 to 100. Chen's teacher gives a student a straight A if his or her average is at least 93. (The teacher does not round off grades so a $92\frac{5}{6}$ average would receive an A-). After 4 tests Chen's average was 94. What was the lowest score that Chen could have received on the 5th test if he got an A for the class?

4. (L3) Martin's teacher decided to drop the lowest score when computing grades. Martin was very excited because he was sick during the third test and scored 64 on it. If dropping the 64 raised Martin's average from 88 to 92, how many tests were there in the semester?

5. (L3) Twenty students took a math test. All scores were whole numbers from 0 to 100. The class average was 97. Given this information, what is the largest possible value for the probability that a student selected at random will have a score that is a multiple of 11?

6. (L4) Seventy percent of a group of Martians are men. Half of them are green. Twenty percent of the group are women. Sixty percent of the women are green. The other ten percent if the group are neither male nor female. What is the smallest possible number of Martians in the group if there are eight more red Martians than green Martians?

Meet #5 Algebra: 5.1 Solving Quadratic Equations

1. (L1) Find the set of all solutions to $x^2 - 5x + 6 = 0$.

2. (L1) Let r and s be the solutions to $x^2 - 10x + 16 = 0$. Find $r + s + rs$.

3. (L2) Find the largest solution to $2x^2 - 6x + 3 = 0$. Give your answer as a reduced fraction in simplest radical form.

4. (L2) What is the difference between the largest and smallest solutions to the equation $2x^2 + 13x - 7 = 0$?

5. (L3) Find all solutions to $x^2 + 10x - 999 = 0$.

6. (L3) Find the set of all x for which $(x + 1)^4 - 5(x + 1)^2 + 4 = 0$

7. (L4) For how many integers b does the quadratic equation $2x^2 + bx - 60 = 0$ have two integer solutions?

Meet #5 Algebra: 5.2 Graphs of Quadratic Equations

1. (L1) The graph of the quadratic equation $y = x^2 - ax + b$ passes through the points (5, 0), and (8, 0). Find the ordered pair (a, b)

2. (L2) The quadratic equation $y = ax^2 + bx + c = 0$ is graphed below. For what value of x does this quadratic take on its least possible value?

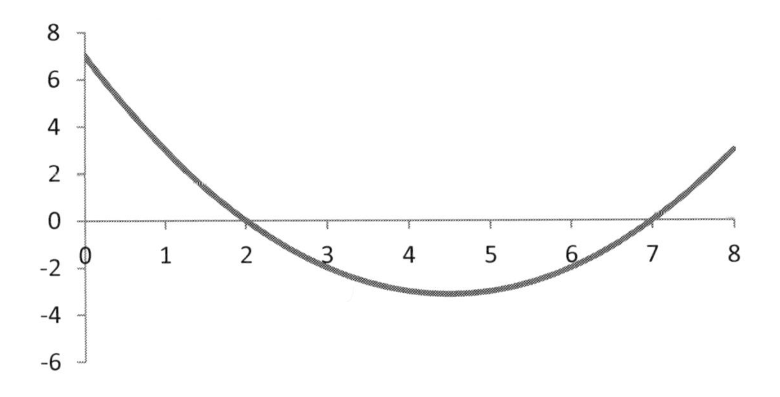

3. (L3) The graph of the quadratic equation $y = ax^2 + bx + c$ passes through the points (0, –6), (3, 0), (6, 0), and (7, z). What is z?

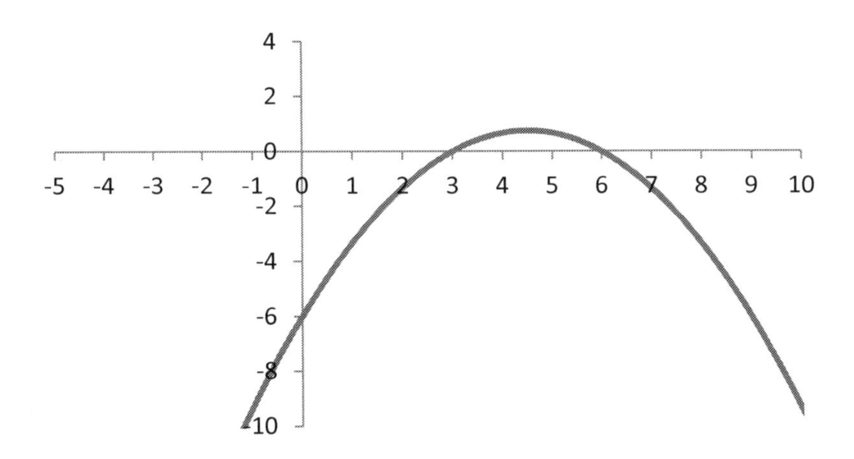

Meet #5 Algebra: 5.3 Problems That Involve Quadratic Equations

1. (L1) Find x if x is positive and $x^2 - 4x - 18 = 3$.

2. (L1) What is the value of x if the triangle shown below is a right triangle?

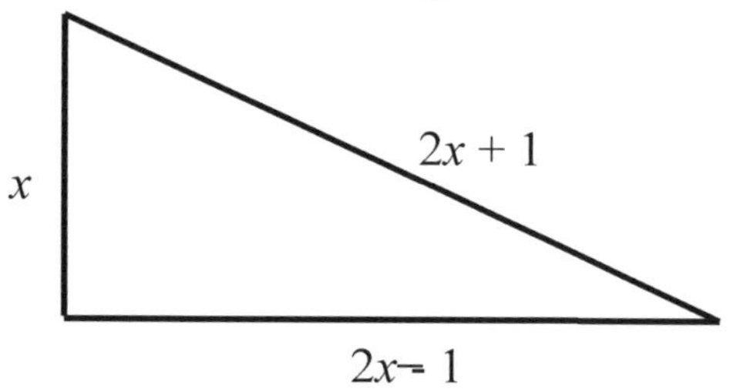

3. (L2) The square of Kate's age is forty nine less than the product of her grandmother's age and her dog's age. How old is Kate if her grandmother is 74 years older than her and her dog is two years old?

4. (L2) One edge of a rectangle is 2cm more than 4 times the length of another edge. What is the perimeter of the rectangle if its area is 12cm^2?

5. (L3) Max takes his pig and monkey with him while walking to the mailbox. The pig runs back home at a constant rate. Max walks back at a rate that is 3 ft/second slower. The monkey starts with Max, but then runs ahead, catches the pig, and rides the rest of the way on the pig's back. The mailbox is 720 feet from Max's house. The monkey arrives home 54 seconds before Max does. (The monkey doesn't have a smartphone and therefore cannot use this time to watch "Baby Monkey Riding Backwards on a Pig" on YouTube.) How fast does Max walk in feet per second?

6. (L3) The difference between the reciprocal of a number and the reciprocal of one less than four times the number is equal to 10. What is sum of the reciprocals of the two possible values for the number?

Meet #5 Algebra: 5.4 Advanced Topic: Complex Numbers

1. (L1) The solutions to the quadratic equation $x^2 - 2x + 5 = 0$ can be written as $a + bi$ and $a - bi$ where a and b are positive integers and $i = \sqrt{-1}$. Find (a, b).

2. (L1) The solutions to the quadratic equation $x^2 + bx + c$ are $3 + 2i$ and $3 - 2i$. Find (b, c).

3. (L2) Let r and s be the solutions to $x^2 - 11x + 24 = 0$. Find $r^2s + s^2r$.

4. (L2) Let r and s be the solutions to $x^2 + 13x + 72 = 0$. Find $\frac{1}{r} + \frac{1}{s}$.

5. (L3) Let r and s be the solutions to $x^2 + 17x + 97 = 0$. Find $r^2 + s^2$.

6. (L3) Let $i = \sqrt{-1}$. The expression $i^3 - 2i^2 + 3i - 2$ can be written in the form $a + bi$ where a and b are real numbers. Find (a, b).

7. (L4) Let $i = \sqrt{-1}$. Write $(2 + i)^2 - \frac{1}{2+i}$ in the form $a + bi$ where a and b are real numbers.

Mathcounts® Geometry: 2.1 Special Triangles

1. (L1) ABC is an isosceles right triangle with AB=2 cm. AD is perpendicular to BC. Find AD in cm in simplest radical form.

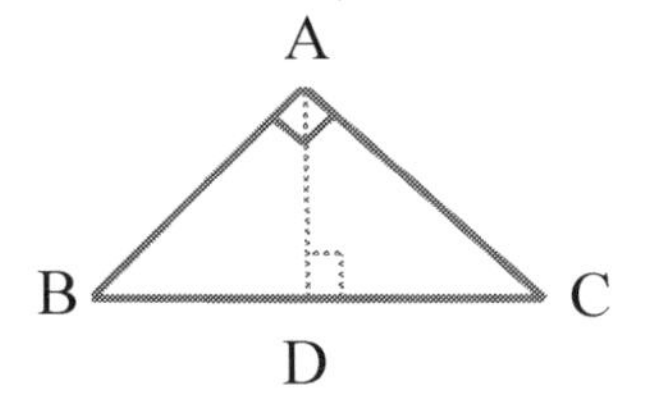

2. (L2) Let ABC be an equilateral triangle with side length 1. Let O be a point in the interior of the triangle satisfying OA=OB=OC. Extend AO to meet side BC at point D. What is the length of OD in simplest radical form?

3. (L2) In rectangle ABCD, point E is located on side AD with AE=AB and CE=2 DE. What is the ratio of ED to EA in simplest radical form?

4. (L3) Let ABC be a triangle with sides of length 2, 2, and $2\sqrt{2}$. Let O be the circumcenter of the triangle and let G be the incenter of the triangle. Find the square of the distance between O and G in simplest radical form.

5. (L3) In the figure on the left below, the squares are congruent and have sides parallel to sides of the regular octagon with side length 1. What is the length of the sides of the squares? Give your answer in the form $\frac{a+\sqrt{b}}{c}$.

6. (L4) In the figure on the right above the measures of CAB, ABC, and BCA are 60°, 45°, and 75°, respectively. Point D on AB is such that ΔACD is similar to ΔABC. What is DB/AC? Give your answer in the form $\frac{a+\sqrt{b}}{c}$.

Mathcounts® Geometry: 2.2 Analytic Geometry: Lines (pt 1)

1. (L1) The line $y = ax + b$ passes through the points (0, 5) and (10, -8). What is a? Express your answer as a decimal.

2. (L1) The line $ax + by + c = 0$ passes through the points (4, 12) and (10, 6). Find (a, b, c) if a, b, and c are integers with GCF (a, b, c)=1 and a is positive.

3. (L2) The line $y = ax + b$ passes through the point (3, 17), has a slope that is a prime number, and intersects the positive y axis. What is the smallest possible value of b?

4. (L2) The vertices of triangle ABC are the points of intersection of the lines $y = 0$, $y = 3x + 6$, and $y = 4 – 2x$. What is the area of ABC? Express your answer as a decimal.

5. (L3) What is the distance between the point of intersection of the lines $3x + 5y = 9$ and $y = 3x + 9$ and the positive x axis? Express your answer in simplest radical form.

6. (L3) Let ABCD be a trapezoid with AB = BC = CD = 10 and AD = 20. Let E be the midpoint of BC and let F be the point where segments AC and DE intersect. Find the area of triangle AFD. Express your answer in simplest radical form.

7. (L3) The line $y = mx + b$ passes through the points (2, 3) and (b, m). What is product of all possible values for m?

Mathcounts® Geometry: 2.2 Analytic Geometry: Lines (pt 2)

1. (L1) What is the midpoint of the segment connecting (1, 0) and (2, 5)? Express your answer as an ordered pair of common fractions.

2. (L1) What is the distance from the point (6, 2) to the line y = 10 – 3x? Express your answer in simplest radical form.

3. (L2) The "median" of trapezoid ABCD is defined to be the segment connecting the two nonparallel sides half way in between the two bases. For example, if the trapezoid has vertices A (-2, 0), B (2, 6), C (10, 6), and D (11, 0), then the median is the segment EF where E is the midpoint of AB and F is the midpoint of CD. In this example, what are the coordinates of the point on the median that is twice as far from E as from F?

4. (L2) Let ABC be a triangle with AB=5, BC=12, and AC=13. Let D be the midpoint of AC. What is the distance from B to D? Express your answer as a common fraction.

5. (L2) Let ABC be the triangle with vertices A (0, 0), B(10, 0), and C (0, 12). The "circumcenter" O of a triangle is defined to be the point that is equidistant from all three points of the triangle. (The name reflects that there is a circle centered at O that passes through all three vertices.) What are the coordinates of the circumcenter of this triangle?

6. (L3) The set of all points equidistant from the points (1, 3) and (5, 10) can be written in the form y = ax + b. What is b? Express your answer as a mixed number.

7. (L4) A pyramid is formed by cutting off the corner of a unit cube along a plane that intersects each edge one-third of the way from the vertex to each adjacent vertex. If the pyramid is placed on a table top in such a way as to make its height as small as possible, what is its height? Express your answer in the form $\frac{a\sqrt{b}}{c}$ where a and c have no common factors and b has no square factors.

Mathcounts® Geometry: 2.3 Analytic Geometry: Circles

1. (L1) What is the area of the region satisfying $(x-1)^2 + y^2 \leq 9$?

2. (L2) The circle $(x-3)^2 + y^2 = r^2$ is tangent to the line $y=3x$. What is the y coordinate of the point of tangency?

3. (L2) The circle $(x-a)^2 + y^2 = r^2$ passes through the points (1,0), (3, 4), and (10,-3). What is r?

4. (L3) Suppose A and B are points on the circle $x^2 + y^2 = 9$ with AB=3. Let C be a point on the line connecting A to the origin which is such that the line segment CB is tangent to the circle. Find the measure of angle ACB in degrees.

5. (L3) What is the area of the region in the first quadrant satisfying $x^2 + 4x + y^2 \leq 12$.

6. (L4) The circle $x^2 - ax + y^2 - 2y + b = 0$ is tangent both to the line $7x - 6y = 0$ and to the line $2x + 9y - 8 = 0$. What is the value of a?

7. (L4) Find the area of the region defined by $x^2 + y^2 \leq 3 + |2x|$?

Mathcounts® Geometry: 2.4 Areas of Polygons on a Grid

1. (L1) How many points with integer coordinates are in the interior of the pentagon bounded by (0, 0), (10, 0), (10, 4), (7, 6), and (0, 6)?

2. (L2) What is the area of the convex octagon with vertices at all points of the form (±3, ±5) and (±5, ±3).

3. (L2) What is the area of the convex pentagon with vertices at (1, 1), (-3, 2), (5, -1), (-2, -3), and (1, -1)?

4. (L3) How many points with integer coefficients are interior to the convex hexagon with vertices at all points of the form (±3, ±9) and (±9, 0) and exterior to the convex hexagon with vertices at all points of the form (±1, ±3) and (±3, 0)?

5. (L3) What is the area of the set of all points (x, y) that satisfy $3x + 2y \leq 14$, $x - 2y \geq -6$, and $x + 6y \geq -6$?

6. (L4) How many ordered pairs of integers (x, y) satisfy the equations $6 < |x - 3| + 2|y - 5| < 10$ and $|xy| > 0$?

Mathcounts® Geometry: 2.5 The Triangle Inequality

1. (L1) The sides of a triangle have lengths 5 cm, 6 cm, and x cm. How many different integer values for x are possible?

2. (L2) What is the smallest possible area for a triangle in square inches if the lengths of the sides (in inches) are consecutive integers. Write your answer as a decimal to the nearest tenth.

3. (L2) The lengths of the sides of a triangle (when measured in inches) are distinct prime numbers. What is the smallest possible value for the perimeter of the triangle?

4. (L3) The sides of a triangle have lengths 5, x, and y. How many different ordered pairs (x, y) are possible if x and y are whole numbers with $x+y \leq 10$.

5. (L3) Let S be a set of triangles. Suppose that all triangles in S have side lengths that are one digit prime numbers and that no two triangles in S are similar. What is the largest possible number of elements of S?

6. (L4) The sides of a triangle have lengths x+y, 2x-y, and 2x. What is the largest possible value for y/x if the length of each side is a one digit whole number? Give your answer as a common fraction.

Mathcounts® Geometry: 2.6 A Few Random Facts

1. (L1) Find the area of the rhombus with vertices (-1, -1), (0, 4), (4, 0), and (5, 5).

2. (L2) Let ABC be a triangle with $\overline{AB} = \overline{AC} = 8$cm and with the measure of angle ABC being 30°. What is the area of ABC in cm^2?

3. (L2) A regular octagon is inscribed in a circle of radius 1. What is the area of the octagon? Express your answer in simplest radical form.

4. (L3) What is the largest possible value for the area of a parallelogram if the lengths of its diagonals are 13 cm and 17 cm?

5. (L4) Suppose that the measure of OA is 15 units. Let B and C be points on the circle of radius 8 units centered at O which are such that the measure of angle BOA is 45 degrees and the measure of minor arc BC is 90 degrees. Find the largest possible value for the absolute value of the difference between the area of triangle BOA and the area of triangle COA.

6. (L4) Let AB be a segment of length 6 and suppose that ABD, ABE, ABF, and ABG are distinct triangles each with side lengths 4, 5, and 6. Suppose AE and BD intersect at X and AF and BG intersect at Y. What is the smallest possible value for the area of AXBY?

Mathcounts® Number Theory: 3.1 More Modular Arithmetic: The Fermat-Euler Theorem

1. (L1) How many of the first 100 counting numbers are neither divisible by 2 nor divisible by 5?

2. (L1) What is the remainder when 7^{49} is divided by 65?

3. (L2) What is the units digit when 123^{321} is written out as a base ten number?

4. (L2) What is the tens digit when 1449^{124} is written out as a base ten number?

5. (L3) What is the remainder when $2015^{(7^{13})}$ is divided by 7?

6. (L3) What is the remainder when 343^{14} is divided by 100?

7. (L3) What is the remainder when 5^{121} is divided by 65?

8. (L4) If $\lfloor x \rfloor$ means the largest integer less than or equal to x, what is $\frac{111^{2002}}{1000} - \left\lfloor \frac{111^{2002}}{1000} \right\rfloor$?

9. (L4) How many two digit positive integers x satisfy GCF(x, 90) = GCF(x, x+11)?

Mathcounts® Number Theory: 3.2 Repeating Decimals

1. (L1) What is the 99th digit after the decimal point in the decimal expansion for 12/17?

2. (L2) What is the 100th digit after the decimal point in the decimal expansion for 1/77?

3. (L2) What is the 50th digit after the decimal point when 4/35 is written as a decimal?

4. (L3) What is the smallest positive integer k for which $10^k \equiv 1 \pmod{53}$?

5. (L3) What is the 18th digit after the decimal point when 6/53 is written as a decimal?

6. (L4) What is the smallest positive integer k for which the 100th digit after the decimal point in the decimal expansion of $1/(5^k \cdot 37)$ is a 1?

7. (L4) Let S={11, 1221, 123321, 12344321, …, 12345678987654321}. Let n be the smallest element of S for which the repeat length for 1/n is not a factor of 12. What is n?

8. (L4) Numbers that are not whole numbers can be written in the base 8 number system just like they can be written as decimals in the base 10 number system. For example, $a.bcd_{(8)}$ means $a + \frac{b}{8} + \frac{c}{8^2} + \frac{d}{8^3}$. Suppose that the fraction 5/9 is written in the base 8 number system. What is the 50th digit after the octimal point?

Mathcounts® Arithmetic: 4.1 Series and Products

1. (L1) The first term of an arithmetic sequence is 7 and the tenth term is 28. What is the one hundredth term?

2. (L1) What is the sum of all positive integers that are less than one hundred and give a remainder of 3 when divided by 7?

3. (L2) The MathCounts Handbook contains 300 problems. In the first week Sophie does 57 of the problems. In each of the next four weeks she does one-third of the problems that remain undone at the start of the week. How many problems does she do in the first five weeks?

4. (L2) Find the sum of the numbers in the matrix below:

$$\begin{matrix} 1 & 2 & 3 & \dots & 100 \\ 3 & 4 & 5 & \dots & 102 \\ \dots \\ \dots \\ 101 & 102 & & \dots & 200 \end{matrix}$$

5. (L3) The first term of a sequence is 191 and every term after the first is obtained by multiplying the square of the tens digit of the previous term by the sum of the digits of the previous term. For example, the second term is $81 \times 11 = 891$. What is the 2017th term in the sequence?

6. (L4) Simplify $\frac{1}{1}\cdot\frac{1}{3}+\frac{1}{3}\cdot\frac{1}{5}+\cdots+\frac{1}{2015}\cdot\frac{1}{2017}$ and express as a fraction.

Mathcounts® Arithmetic: 4.2 Review of Standard Combinations Problems

1. (L1) There are 10 kids on a Mathcounts team. In how many ways can the coach choose four to be the school team and six to compete as individuals?

2. (L2) John has six math books on his shelf. On January 1st, 2017 he makes a New Year's resolution. At 5pm that day and on every day thereafter he will rearrange the books and put them in an order that have never been in before. On what date will he have to break his resolution?

3. (L2) In how many ways can 10 kids be split up into two teams of 5?

4. (L2) Miss Laizee teaches 102 students. Every day she picks two or three to be homework helpers. (They stay after school and grade the day's homework.) In how many ways can she choose the homework helpers?

5. (L3) For how many values of N is the number of ways to choose a 4 element subset from an N element set a prime number?

6. (L3) Mr. Yellin wants to divide 10 kids into two teams of 5 for a practice debate. In how many ways can she do this if she wants to make sure that her two best debaters, Amy and Elizabeth, are on opposite teams?

7. (L3) Humberto's math teacher picks three students to come up in front of the class. Humberto is bored, so he decides to figure out that the number of ways in which the three students could have been chosen. If he correctly figures that the answer is 1330, how many students are in the class?

Mathcounts® Arithmetic: 4.3 Combinations Problems with Three or More Groups

1. (L1) A quidditch team has seven players. They need to choose one person to be their seeker, 3 to be chasers, 2 to be beaters, and 1to be the keeper. In how many ways can this be done?

2. (L1) There are 12 students on a math team. In how many ways can the coach choose 4 students to be his official Mathcounts team and 6 students to be individual competitors?

3. (L2) There are 10 kids on the Bigelow Mathcounts team. Four (including Anna) will go to the meet in Mr. Ellison's car. Four (including Julia) will go in Julia's mother's car. The other two will go with another parent. Julia decides that the fairest way to decide who will go in each car would be to make a list with all of the possible arrangements and ask kids to rank them from their most to least favorite writing 1 for their favorite, 2 for their second favorite, …, and N for their least favorite. What will N be?

4. (L3) Min Jae's chessboard has 64 squares. The rows are labeled 1, 2, …, 8, and the columns are labeled A, B, …, H. In how many distinguishable ways can he place 3 identical white pawns and the white king on the board?

5. (L4) A science class has 22 students. The teacher wants to divide the class into groups with 4 or 5 students each to do a project. In how many ways can this be done?

6. (L4) Three times during the year the coach of a ten student math team has a pair of tickets to give away and decides to do this by having a within –team contest and giving the tickets to the two highest scorers. Once the pair is to a Red Sox game, once it is to see Meghan Trainor, and once it is to see Yo Yo Ma. In how many of the possible ways that the tickets may be given out does someone go to more than one event?

Mathcounts® Arithmetic: 4.4 Using Combinations to Count Orderings

1. (L1) How many different 5-digit numbers have two 2's and three 3's as digits?

2. (L1) How many different 7-letter sequences can be formed using seven letter tiles that spell out REORDER?

3. (L2) In how many ways can the letters in the word MIDDLE be arranged if the two D's must be kept together?

4. (L2) How many 6-letter palindromes can be formed by arranging letter tiles that spell out HANNAH?

5. (L3) How many of the 5-digit numbers that can be formed by rearranging the digits in the number 11,234 have the 1's separated?

6. (L3) How many 8-digit numbers can be formed using the digits 0, 2, 3, 3, 3, 5, 6, and 7?

7. (L3) How many palindromes can be made by rearranging the letters ABBABBACC?

Mathcounts® Arithmetic: 4.5 Counting Ordered Ways to Make Numbers Add Up

1. (L1) Ting Han rolls a six-sided die three times. The numbers he gets add up to eight. How many sequences could possibly be the sequence of rolls?

2. (L1) In how many ways can 12 be written as a sum of 3 nonnegative integers? Count $8 + 0 + 4$ and $0 + 4 + 8$ as two different ways.

3. (L2) In how many ways can 12 be written as an ordered sum of three one-digit positive numbers?

4. (L2) How many paths from A to B along the grid on the left below have length seven?

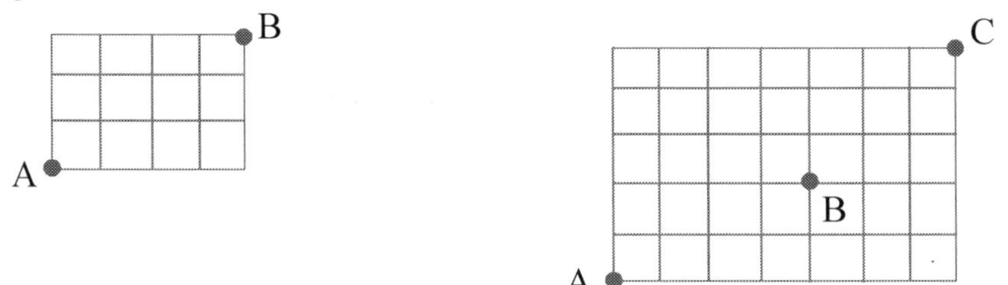

5. (L3) How many paths from A to C on the grid on the right above have length 12 and pass through the point B?

6. (L3) How many whole numbers x satisfy all of the following?

- x is a 5-digit positive whole number.
- x is less than 20,000.
- The digits of x are in strictly increasing order.
- x is one less than a multiple of 10.

(For example, one such number is 12,569.)

7. (L4) If a way to write 9 as an ordered sum of two or more positive integers is chosen at random with all such sums being equally likely, what is the probability that all numbers in the chosen sum are odd?

Mathcounts® Arithmetic: 4.6 Counting Problems Without an Easy Answer: Organize Your List

1. (L1) How many three element subsets of {1, 2, 3, 4, 5, 6, 7} have one element equal to the sum of the other two elements?

2. (L2) In how many ways can 12 be written as a sum of (not necessarily distinct) prime numbers? Count different orderings of the same sum as different ways.

3. (L2) How many different products can be obtained by multiplying two different numbers from the set {-8, -2, -1, 0, 1, 4}?

4. (L2) How many squares can be formed using points from a regular 3×4 grid as vertices?

5. (L3) Let S be a set of triangles no two of which are congruent. Suppose that all triangles in S have side lengths that are whole numbers less than 25. Suppose that in each triangle the product of the side lengths is a multiple of 49. What is the largest number of elements that S could have?

6. (L4) Sicong noticed that the 6-digit number that you get when you write the first six Fibonnaci numbers next to each other, 112,358, has several interesting properties: each digit is at least as big as the one that comes before; it starts and ends with a power of 2; and one digit (the "1") occurs twice in row. How many 6-digit numbers have all of these properties?

Mathcounts® Arithmetic: 4.7 Probability Problems Involving Combinations

1. (L1) If a two-element subset of {1, 2, …, 20} is chosen at random, what is the probability that the subset is a pair of twin primes?

2. (L1) A bag has 14 silver and 3 gold balls. If you choose two balls at random from the bag (without replacement), what is the probability that you will get one ball of each type?

3. (L2) The Day math team has 6 boys and 4 girls. If they choose a Mathcounts team by picking four students at random, what is the probability that their team will have exactly one girl?

4. (L2) In the Powerball lottery 5 distinct white balls are drawn from a set numbered 1 to 69 and 1 red ball is chosen from a set numbered 1 to 26. To win you need to guess the 5 white numbers and the 1 red number (order doesn't matter within the white balls). What is the probability of winning?

5. (L3) A bag contains 9 red balls, 5 white balls, 5 blue balls, and one green ball. If four balls are taken out of the bag at random, what is the probability that the four balls taken out have at least two balls of some color?

6. (L3) Scrabble tiles that spell out VIKRAM KRISHNAMACHARI are placed in a bag. Six tiles are chosen randomly from the bag. What is the probability that the six tiles can be rearranged to spell VIKRAM?

7. (L4) We say that name has a double letter if the same letter occurs two times in a row. For example, CONNOR has a double letter because the third and fourth letters are both N, but GREGORY does not have a double letter because the G's and R's are both separated. Take the name CONNOR and rearrange the letters at random. What is the probability that the resulting name has a double letter?

Mathcounts® Arithmetic: 4.8 Binomial Problems and Pascal's Triangle

1. (L1) A fair coin is tossed five times. What is the probability that heads will come up an odd number of times?

2. (L1) If a fair coin is tossed seven times what is the probability that heads will come up two or fewer times?

3. (L2) Every time Ted comes to bat there is a 40% chance that he will get a hit. What is the probability that Ted gets exactly two hits in five at bats? Express your answer as a decimal.

4. (L2) A fair coin is flipped five times. What is the probability that the first toss was heads if at least two of the five tosses came up heads?

5. (L3) Two fair coins are flipped six times each. What is the probability that each coin comes up heads the same number of times?

6. (L3) Five six-sided dice are rolled. What is the probability that at least three dice have a number of at least three on the top if there is at least one die with a number of at least three on top?

7. (L4) Three six-sided dice are rolled. What is the largest value of k for which the probability that the median of the numbers on the three dice is at least k is at least one-half?

8. (L4) Ten twenty-sided dice are rolled. What is the probability that exactly five of the numbers that come up are prime and exactly three are composite?

Mathcounts® Algebra: 5.1 Proportional Reasoning

1. (L1) Pyramid A has a volume of 72 cm^3 and its base is a square with side length 9cm. Pyramid B has the same height as pyramid A and its base is a square with side length 6cm. What is the volume of pyramid B in cm^3?

2. (L1) Triangle ABC has an area of 60cm^3 and $\overline{BC} = 18$ cm. If D and E are the midpoints of AB and AC, respectively, what is the ratio of the area of ADE to the area of trapezoid DECB? Give your answer as a fraction.

3. (L2) Johann paid $0.92 in sales tax on a $16 item. If a $40 item is taxed at the same rate how much would the sales tax be?

4. (L2) Five identical robots working together can paint 5 rooms in 5 days. How long would it take 3 of the robots to paint 9 rooms?

5. (L2) A pool holds 1440 cubic feet of water. A large hose fills the pool in 4 hours. If the large lose and a small hose are both turned on, the pool will fill in 3 hours. How long would it take to fill the pool with the small hose?

6. (L3) Cathy started biking to Vivian's house at 10mph. When she was halfway there she got a flat tire, left her bike, and walked the rest of the way at 4mph. In total she took 40 minutes to get to there. At 5 pm Vivian's sister drove Cathy to her bike at 25 mph and they spent 5 minutes changing the tire. Vivian then biked home at 10 mph. When did she get home?

7. (L3) Kevin runs around his school's 400m track at 80 seconds per lap. He and Elena start running at the same time from the same place, but run around the track in the opposite direction. If they pass each other for the 10th time exactly 8 minutes after they started, what was Elena's average speed in meters per second? Give your answer as a common fraction.

Mathcounts® Algebra: 5.2 Polynomial Division with Remainders

1. (L1) Use polynomial division to find the remainder when $2x^2 + 6x + 3$ is divided by $x + 2$.

2. (L1) For what value of a is $x + 3$ a factor of $x^3 + 13x^2 - 14x + a$?

3. (L2) Suppose $5x^2 + 7x + 3 = (ax + b)(x - 3) + c$ for all x, where a, b, and c are whole numbers. What is c?

4. (L2) What is the value of $x^3 - 5x^2 + 9x + 8$ if $x^2 - 2x + 3 = 0$?

5. (L3) The function $f(x) = x^3 - 9x^2 + 23x - 15$ satisfies $f(3)=0$. What is the largest value of x for which $f(x)=0$?

6. (L3) The area in square inches of square ABCD is one more than one half of its perimeter in inches. What is the value of the product $\overline{AB} \cdot \overline{AC} \cdot \overline{AD}$, where $\overline{AB}$ denotes the distance between points A and B? Give your answer in the form $a + b\sqrt{c}$ where a, b, and c are whole numbers and c is square-free.

7. (L4) What is the sum of all possible values for $6x^3 + 13x^2 - 15x + 75$ if $2x^2 = 5x + 7$? Give your answer as a common fraction.

Mathcounts® Algebra: 5.3 Systems of Equations

1. (L1) Find the point of intersection of the lines $3x - 2y = 5$ and $y = 4x + 2$. Write your answer as an ordered pair of decimals.

2. (L1) In pentagon ABCDE angles EAB and ABC are each twenty degrees larger than angle BCD and angles BCD and CDE are each twenty degrees larger than angle DEA. What is the measure of angle EAB in degrees?

3. (L2) Find the sum of all two-digit numbers that have the property that they are 600% larger than the sum of their digits.

4. (L2) The sum of three more than twice a number and one less than a second number is 5. The difference between twice the first number and the second number is 4. What is the absolute value of the larger of the two numbers? Express your answer as a common fraction.

5. (L3) Anna and Bob each walked in the Walk for Harder Math. Anna started an hour after Bob did. She walked two km/hr faster than Bob and finished thirty minutes earlier than he did. If Anna had walked one km/hr faster than she did, then she would have finished one hour before Bob. How many kilometers long was the Walk for Harder Math?

6. (L3) Julia's store sells three products: bingles, bangles, and bongles. A customer buys 23 items for a total of $87. A bingle costs twice as much as a bongle. A bangle costs as much as two bingles and one bongle. The number of bingles that the customer bought is five less than the number of bangles. The number of bongles that the customer bought is four more than the number of bangles. How much does a bangle cost?

Mathcounts® Algebra: 5.4 Maximizing Quadratic Functions

1. (L1) What is the largest possible value for $6x - x^2$ if x is a real number?

2. (L1) What is the largest possible value for the product of two real numbers if their sum is 17?

3. (L2) There is a long straight cliff at the edge of John's back yard. He thinks it would be cool to have a rectangular skating rink with one edge right along the cliff. The town inspector thinks this is a terrible idea and tells John that he would require that he put a fence on the other three sides to keep neighborhood children from sneaking onto the rink and falling to their deaths. If John has enough money to buy 200 feet of fencing, what is the area in square feet of the largest rink he can build?

4. (L2) The formula for converting temperatures in Celsius to Fahrenheit is $F = \frac{9}{5}C + 32$. What is the smallest possible value for the product of the temperature in degrees Fahrenheit and the temperature in degrees Celsius?

5. (L3) Kara gets $43 - 3x$ likes on each photo if she posts x photos to Instagram in one day. How many photos should she post in a day if she wants to get the largest possible total number of likes?

6. (L3) What is the smallest possible value for $(x^2 - 37)(x^2 + 11)$ if x is an integer?

7. (L4) Let ABC be a right angle. Let D and E be points on AB and AC with DE parallel to BC. Let G be a point on DE with AFGD a square. Suppose $\overline{AB} = 1$ cm, $\overline{DE} = x$ cm, and $\overline{BC} = y$ cm. Find the ordered pair (x, y) if the length of DF in cm exceeds the area of AFGD in cm^2 by the largest amount that is possible when x and y are one-digit whole numbers?

Answers

Meet #1 Geometry

2.1: Basic Definitions
1. 120°
2. 20°
3. 35°
4. 90°

2.2: Adding Up Rules
1. 25°
2. 15°
3. 20°
4. 127°
5. 138.6°.

2.3: Equality Rules
1. 140°
2. 62°
3. 132°

2.4: Angles in Polygons
1. 60°
2. 144°
3. 12
4. 74°

2.5: Problem Solving Strategies
1. 18°
2. 51.5927°
3. 115°.

2.6: Advanced Topics
1. 36°
2. 15°
3. 97.5°

Meet #1 Number Theory

3.1: Prime Numbers
1. 23, 29
2. 37
3. 29, 31
4. $40\frac{1}{14}$
5. 10
6. 91

3.2: Prime Factorization
1. 13
2. 2
3. 5
4. 6
5. 46
6. 5
7. 79
8. 24

3.3: Counting and Summing Factors
1. 1, 3, 5, 15, 25, 75
2. 9
3. 36
4. 465
5. 45
6. 2
7. 240
8. 738150

3.4: Divisibility Rules
1. 0, 5
2. 2
3. 0
4. 4
5. 12
6. 1
7. (4, 6)
8. 6

3.5: Problem Solving Tips
1. 3
2. 25
3. 1
4. 26
5. 46656
6. 216

Meet #1 Arithmetic

4.1: Order of Operations

1. 0.875
2. 2
3. 1.67
4. 6

4.2: Statistics

1. 600.4
2. 88
3. 1
4. 99
5. 3

Meet #1 Algebra

5.1: Simplifying Expressions

1. 7x – 1
2. 42 – 16x
3. 406x – 1000
4. 2191x – 30
5. 15x + 50
6. 29x – 3

5.2: Evaluating Expressions

1. 11
2. 8
3. 1
4. 6/7
5. $7\frac{1}{17}$
6. 1/7

5.3: Solving Equations in One Unknown

1. 5
2. 1.8
3. 1
4. 79 cents
5. -39/5
6. 14

5.4: Identities

1. 6	2. 3
3. 1/6	4. 3
5. 620	6. 1

5.5: Made-up Operations

1. 1	2. 17
3. 18856	4. -128
5. 16320	6. 261

Meet #2 Geometry

2.1: Perimeter
1. 12 cm
2. $4 + 6\sqrt{5}$ (units)
3. $24 + 18\sqrt{3}$ cm.

2.2: Areas
1. 225 square parsecs
2. $4 + 8\sqrt{2}$ square cm
3. 48 square feet.

2.3: Perimeters of Rectilinear Figures
1. 11,000 feet
2. 68 cm
3. 128 cm

2.4: Problem Solving Tip: Subtract
1. 32 cm^2
2. 336 sq. inches
3. 2¼ sq. feet

2.5: More Area Formulas
1. 84 cm^2
2. $\frac{\sqrt{6}}{2}$ cm
3. 2.83 sq. ft. ($2\sqrt{2}$)
4. 75 sq. feet.

Meet #2 Number Theory

3.1: A Super-Quick Review of Prime Factorization
1. 16
2. {1, 13, 169}
3. 8
4. 5
5. $2^1 \times 5^1 \times 11^2$
6. February 2, 2003.

3.2: Greatest Common Factors
1. 3
2. 7
3. 17,150
4. 3
5. 1 min 37 sec
6. 6
7. 1

3.3: Least Common Multiples
1. 120
2. 110
3. 42,420
4. 3:35:45 pm
5. 30,342
6. 4

3.4: More on GCFs and LCMs
1. 1
2. 336
3. 24,642
4. 11:27
5. 5
6. 98,088.

3.5: A Longer Review of Prime Factorization
1. 320
2. $2^2 \times 3^2 \times 11^1$
3. 12
4. 360
5. 124
6. 1500

Meet #2 Arithmetic

4.1: Fractions and Percents

1. 34
2. 65
3. 9
4. 39.78
5. 2500 pounds
6. 237

4.2: Terminating and Repeating Decimals (pt 1)

1. 5/11
2. 1/6
3. 11/90
4. 7/15
5. 1/1098900
6. $0.00\overline{000091}$

4.2: Terminating and Repeating Decimals (pt 2)

1. $0.0\overline{3}$
2. 6
3. 8
4. $0.\overline{07317}$
5. 3
6. 0
7. 17

Meet #2 Algebra

5.1: Sums of Arithmetic Sequences

1. 495
2. 11
3. 214
4. Tuesday
5. 84
6. -1024

5.2: Reasoning In Number Sentences

1. 2
2. -1/2
3. 5
4. 5
5. 5
6. 87

5.3: Working with Formulas

1. 17
2. 15 inches
3. 2827 cm^2
4. 12
5. 6
6. 5/2

5.4: Word Problems with One Unknown

1. 17
2. 4
3. 12
4. 11
5. 29
6. 4

Meet #3 Geometry

2.1: Angles in Polygons

1. 10
2. 168°
3. 9
4. 7°
5. 60
6. 106

2.2: Areas of Polygons

1. 84 cm^2
2. 48 cm^2
2. 420/29 cm
3. 42 cm
5. 84 cm^2
6. $\frac{1{,}057{,}222{,}656}{17{,}850{,}625}$ cm^2

2.3: Diagonals in Polygons

1. 9
2. 4850
3. 14
4. 16
5. 9
6. 1400
7. $\pi\left(1+\frac{\sqrt{2}}{2}\right)$ sq. cm

2.4: Pythagorean Theorem (part 1)

1. 10 cm
2. 11 cm
3. 30 sq. cm
4. 5 sq. cm
5. 36

2.4: Pythagorean Theorem (part 2)

6. 5
7. 7 cm
8. 45 feet
9. 102 cm

2.4: Pythagorean Theorem (part 3)

10. 5.5 cm
11. 1.3 miles
12. $99\sqrt{5}$ cm^2
13. 52 cm

Meet #3 Number Theory

3.1: Scientific Notation

1. 5.4×10^1, 3.28×10^2, 1.63×10^5, 4×10^{-2}, 2.5×10^{-3}, 1.23×10^{-4}
2. 1.23×10^7, 2.43×10^{-3}, 8×10^0, 3×10^5, 2.4×10^{-8}, 8×10^{-5}
3. 2.4×10^5, 4.8×10^{10}, 1.148×10^{-1}
4. 164, .0625
5. 4×10^2, 4×10^{-3}, 6×10^{-9}

3.2: IMLEM Scientific Notation Problems

1. 8×10^2
2. 3.6×10^{-5}
3. 1×10^3
4. 1.7×10^{-24}
5. 1.34×10^5
6. 8.5×10^{-3}

3.3: Basics of Bases

1. 398
2. $67_{(8)}$
3. $211_{(3)}$
4. 3000
5. 522
6. 7
7. 7
8. 32

3.4: Converting from Base A to Base B

1. $201_{(5)}$
2. $110110_{(2)}$
3. $545_{(8)}$
4. $20021101_{(3)}$
5. $121530_{(8)}$
6. $13323_{(4)}$

3.5: Arithmetic in Other Bases

1. $2044_{(5)}$
2. $1750_{(8}$
3. $1221_{(9)}$
4. 1087
5. $1004_{(6)}$
6. $2453_{(8)}$
7. 7
8. 16

3.6: Word Problems Related to Bases

1. 207 grams
2. 262
3. 931 $in.^3$
4. 8

3.7: Adding and Subtracting in Scientific Notation

1. 7.4×10^6, 1.21×10^4, 1.33×10^5
2. 4×10^2, 7.96×10^2, 3.59×10^{-5}
3. 1.066×10^{-2}
4. 0.02938
5. 160

Meet #3 Arithmetic

4.1: Basics of Exponents

1. -17
2. 64
3. $0.\overline{4}$
4. 0
5. 5
6. 7

4.2: Operations with Exponents

1. 3200
2. $2\frac{61}{100}$
3. 25
4. 2.6
5. 25/64
6. 64

4.3: Roots

1. 1
2. 6
3. 49
4. 1
5. $\sqrt{50}$
6. $5\sqrt[3]{50}$
7. 2
8. 3

4.4: IMLEM Questions

1. 90
2. 8
3. 5
4. 3
5. 17
6. 17
7. 11

Meet #3 Algebra

5.1: Linear Equations with Absolute Values

1. 8
2. 4, 8
3. 2, 5
4. 16/3
5. 1
6. 3, 5
7. -3
8. 3/2, 7/2

5.2 Working with Inequalities

1. 0
2. $x < -3$
3. 3/2
4. $x < 0$ or $x > 1/3$
5. 37/7
6. -51
7. -1, 9

5.3 Working with Absolute Values and Inequalities

1. 13
2. 0
3. 5
4. $17\frac{1}{2}$
5. $6\frac{1}{2}$
6. 3
7. 91
8. 3

Meet #4 Geometry

2.1: Areas and Perimeters of Circles

1. 16π cm^2
2. 200
3. $32 - 8\pi$ cm^2
4. 314 mm
5. 20.1 cm^2.

2.2: Arcs and Angles

1. 60°
2. 8π cm^2
3. 50°
4. 35°
5. 16

2.3: A Quick Review of Triangles and Polygons

1. 45°
2. 4
3. 144°
4. 12 cm
5. 6½
6. $\frac{21\sqrt{3}}{32}$ cm^2

2.4: Advanced Circle Facts

1. 20°
2. 1 cm
3. 25/4
4. $2\sqrt{6}$
5. $\frac{35\sqrt{6}}{24}$ cm
6. 30°

Meet #4 Number Theory

3.1: Basic Sequences and Series

1. 83
2. 1050
3. -45
4. 1229
5. 5738
6. 403
7. 15
8. 29

3.2: Modular Arithmetic (pt 1)

1. 2:30am
2. Team 1
3. 1
4. 3
5. 6
6. 8
7. 8

3.2: Modular Arithmetic (pt 2)

1. 3
2. 10
3. 0
4. 55
5. 297
6. 25

3.3: Advanced Sequences and Series

1. 55
2. 16 feet
3. 1540
4. 2021
5. 10101
6. 1⅓ sq. cm
7. 400 feet

Meet #4 Arithmetic

4.1: Percent Applications

1. 61
2. $62.30
3. 20
4. 121
5. $500
6. 11
7. 10.4

4.2: Compound Interest

1. $49612.50
2. $1800.94
3. 27.1
4. 5.06%
6. $40,000
7. $9.03

Meet #4 Algebra

5.1: Functions

1. $5.74
2. $41.56
3. $42.49
4. $40,000
5. 328

5.2: Two Equations in Two Unknowns

1. 33
2. 13
3. 19
4. 1.4
5. 33
6. 8cm^2

5.3: N Equations in N Unknowns

1. 13
2. 47
3. 68
4. $3.94
5. 80 cents
6. 346.5

5.4 Time-Distance-Speed Problems

1. 5 minutes
2. 22.5
3. 2 pm
4. 8.125 mph
5. 5 miles
6. 60 km

Meet #5 Geometry

2.1: Surface Areas and Volumes

1. 27
2. 294 ft^2
3. 480
4. 181.5
5. 152
6. 5

2.2: Shapes Related to Circles

1. 144π sq. in.
2. 90π cm^2
3. 54π
4. 13
5. 0.33
6. $\frac{77}{3}\pi$ cm^2

2.3: Prisms and Pyramids

1. 2,645,000
2. 484
3. 30
4. 192
5. $\frac{200\sqrt{6}}{3}$
6. 4500

2.4: Polyhedra

1. 4
2. 12
3. 40
4. V=60, E=90, F=32
5. 12
6. 4/3

2.5: Units of Measurement

1. 550
2. 30
3. 0.015
4. 839,000
5. 282
6. 729

2.6: Advanced Topics

1. 228π cm^3
2. 816,945
3. 144π
4. 396π cm^3
5. 900π cm^2

Meet #5 Number Theory

3.1: Venn Diagrams

1. 63
2. 6
3. 161
4. 3

3.2: Basic Set Theory

1. 6
2. 6
3. {3, 5, 13}
4. {13, 23, 43, 53, 73, 83}
5. 605

3.3: Inclusion-Exclusion Counting

1. 11
2. 18
3. 63
4. 252
5. 30
6. 71
7. 9
8. 28

3.4: Review of Modular Arithmetic

1. 200
2. 101
3. 67
4. 16
5. 2
6. 30

3.5: Venn Diagrams and Inclusion-Exclusion with Four or More Sets

1. 26
2. 12
3. 1560

Meet #5 Arithmetic

4.1: Combinatorics (part 1)

1. 120
2. 28
3. 999,900
4. 12
5. 720
6. 25

4.1: Combinatorics (part 2)

1. 128
2. 256
3. 19
4. 2106
5. 19,600

4.2: Basic Probability

1. 1/9
2. 1/20
3. 2/9
4. 5/8
5. 17/36
6. 4/5
7. 7/50

4.3: Probability Problems with Two Events

1. 1/9
2. 1/11
3. 16/25
4. 1/12
5. 7/8
6. 1/42

4.4: Using Conditional Probabilities for Complicated Events

1. 3/4
2. 36 in the grid, 25 crossed out, one circled, 1/11
3. 13/30
4. 5/16
5. 4/99
6. 1030/5151
7. 5/32

4.5: Averages

1. 87
2. 3
3. 82
4. 7
5. 19/495
6. 200

Meet #5 Algebra

5.1: Solving Quadratic Equations

1. {2, 3}
2. 26
3. $\frac{3+\sqrt{3}}{2}$
4. 7½
5. -37, 27
6. {-3, -2, 0, 1}
7. 8

5.2: Graphs of Quadratic Equations

1. (13, 40)
2. 4½
3. -4/3

5.3: Problems That Involve Quadratic Equations

1. 7
2. 8
3. 11
4. 19 cm
5. 5 ft/sec
6. 13

5.4 Advanced Topic: Complex Numbers

1. (1, 2)
2. (-6, 13)
3. 264
4. -13/72
5. 95
6. (0, 2)
7. $\frac{13}{5} + \frac{21}{5}i$

Mathcounts® Geometry

2.1: Special Triangles

1. $2\sqrt{2}$ 2. $\sqrt{3}/6$
3. $\sqrt{3}/3$ 4. $6 - 4\sqrt{2}$
5. $\frac{2+\sqrt{2}}{3}$ 6. $\frac{3-\sqrt{3}}{2}$.

2.2: Analytic Geometry: Lines (pt 1)

1. -1.3 2. (1,1,16)
3. 2 4, 9.6
5. $\sqrt{13}$ 6. $40\sqrt{3}$
7. -3/2

2.2: Analytic Geometry: Lines (pt 2)

1. (3/2, 5/2) 2. $\sqrt{10}$
3. (7, 3) 4. 13/2
5. (5, 6) 6. $8\frac{3}{14}$
7. $\sqrt{3}/9$

2.3: Analytic Geometry: Circles

1. 9π 2. 9/10
3. 5/2 4. 30°
5. $\frac{8\pi}{3} - 2\sqrt{3}$ 6. 14/5
7. $\frac{16\pi}{3} + 2\sqrt{3}$

2.4: Areas of Polygons on a Grid

1. 44 2. 92
3. 18 4. 176
5. 20 6. 48

2.5: The Triangle Inequality

1. 9 2. 2.9 $\left(\frac{3\sqrt{15}}{4}\right)$
3. 15 inches 4. 23
5. 11 6. 4/3

2.6: A Few Random Facts

1. 76π 2. 16 cm^2
3. $2\sqrt{2}$ 4. 221/2
5. 0 6. $6\sqrt{7}$

Mathcounts® Number Theory

3.1: More Modular Arithmetic: The Fermat-Euler Theorem

1. 40
2. 7
3. 3
4. 0
5. 6
6. 49
7. 5
8. 321
9. 22

3.2: Repeating Decimals

1. 5
2. 9
3. 1
4. 13
5. 0
6. 8
7. 12344321
8. 5

Mathcounts® Arithmetic

4.1: Series and Products

1. 238 2. 679
3. 252 4. 512,550
5. 81 6. $\frac{1008}{2017}$

4.2: Review of Standard Combinations Problems

1. 210 2. Dec. 21, 2017
3. 126 4. 176,851
5. 1 6. 70
7. 21

4.3: Combinations Problems with Three or More Groups

1. 420 2. 13,860
3. 560 4. 2,541,504
5. 470,531,900 6. 72,225

4.4: Using Combinations to Count Orderings

1. 10 2. 420
3. 120 4. 6
5. 36 6. 5880
7. 12

4.5: Counting Ordered Ways to Make Numbers Add Up

1. 21 2. 91
3. 52 4. 35
5. 300 6. 35
7. 11/85

4.6: Counting Problems Without an Easy Answer: Organize Your List

1. 9 2. 35
3. 10 4. 10
5. 75 6. 649

4.7: Probability Problems Involving Combinations

1. 2/95 2. 21/68
3. 8/21 4. 1/292,201,338
5. 308/323 6. 6/1615
7. 8/15

4.8: Binomial Problems and Pascal's Triangle

1. 1/2 2. 29/128
3. 0.3456 4. 15/26
5. 231/1024 6. 96/121
7. 4 8. 83853/7812500

Mathcounts® Algebra

5.1: Proportional Reasoning

1. 32cm^3 2. 1/3
3. $2.30 4. 15 days
5. 12 hours 6. 5:21pm
7. 10/3 m/sec

5.2: Polynomial Division with Remainders

1. -1 2. -132
3. 69 4. 17
5. 10 6. $10 + 7\sqrt{2}$
7. -17

5.3: Systems of Equations

1. (-1.8, -5.2) 2. 124
3. 210 4. 7/4
5. 36 6. $7.50

5.4: Maximizing Quadratic Functions

1. 9 2. 289/4
3. 5000 sq.ft. 4. -1280/9
5. 7 6. -567
7. (5, 7)

87228231R00072

Made in the USA
Columbia, SC
11 January 2018